NYSTCE 009
CST
Physics
Teacher Certification Exam

By: Sharon Wynne, M.S
Southern Connecticut State University

XAMonline, INC.
Boston

Copyright © 2008 XAMonline, Inc.
All rights reserved. No part of the material protected by this copyright notice may be reproduced or utilized in any form or by any means, electronic or mechanical, including photocopying, recording or by any information storage and retrievable system, without written permission from the copyright holder.

To obtain permission(s) to use the material from this work for any purpose including workshops or seminars, please submit a written request to:

XAMonline, Inc.
21 Orient Ave.
Melrose, MA 02176
Toll Free 1-800-509-4128
Email: info@xamonline.com
Web www.xamonline.com
Fax: 1-781-662-9268

Library of Congress Cataloging-in-Publication Data

Wynne, Sharon A.
    CST Physics 009: Teacher Certification / Sharon A. Wynne. -2nd ed.
    ISBN 978-1-58197-042-5
    1. Physics 009.   2. Study Guides.   3. NYSTCE
    4. Teachers' Certification & Licensure.   5. Careers

Disclaimer:
The opinions expressed in this publication are the sole works of XAMonline and were created independently from the National Education Association, Educational Testing Service, or any State Department of Education, National Evaluation Systems or other testing affiliates.

Between the time of publication and printing, state specific standards as well as testing formats and website information may change that is not included in part or in whole within this product. Sample test questions are developed by XAMonline and reflect similar content as on real tests; however, they are not former tests. XAMonline assembles content that aligns with state standards but makes no claims nor guarantees teacher candidates a passing score. Numerical scores are determined by testing companies such as NES or ETS and then are compared with individual state standards. A passing score varies from state to state.

Printed in the United States of America   œ-1

NYSTCE: CST Physics 009
ISBN: 978-1-58197-042-5

TEACHER CERTIFICATION STUDY GUIDE

Table of Contents

SUBAREA I   **FOUNDATIONS OF SCIENTIFIC INQUIRY**

Competency 1.0   **Understands the historical and contemporary contexts of the study of physics, including the relationships and common themes that connect mathematics, science, and technology and their applications to everyday life**

Skill 1.1   Analyzing the role of early civilizations in establishing the foundations of physics 1

Skill 1.2   Recognizing key events and analyzing the contributions of individuals in the development of physics 2

Skill 1.3   Demonstrating an understanding of current theories of the origin, structure, and composition of the universe 6

Skill 1.4   Assessing the societal implications of physics 7

Skill 1.5   Applying the laws of physics to geological, chemical, biological, and astronomical systems 9

Skill 1.6   Analyzing the use of physics, mathematics, and other sciences in the design of a technological solution to a given problem 9

Skill 1.7   Analyzing the role of technology in the advancement of scientific knowledge 10

Skill 1.8   Evaluating the appropriateness of a variety of software and information technologies to model and solve problems in mathematics, science, and technology 11

Skill 1.9   Analyzing solutions to everyday-life problems that incorporate knowledge and skills of mathematics, science, and technology 13

Competency 2.0   **Understand the process of scientific inquiry and the role of observation and experimentation in explaining natural phenomena**

Skill 2.1   Analyzing processes by which new scientific knowledge and hypotheses are generated 15

Skill 2.2   Analyzing ethical issues related to doing science 19

PHYSICS

TEACHER CERTIFICATION STUDY GUIDE

Skill 2.3 Evaluating the appropriateness of a specified experimental design to test a given physics hypothesis 20

Skill 2.4 Recognizing the role of communication among scientists and between scientists and the public in promoting scientific progress..... 21

Competency 3.0 Understand the processes of gathering, organizing, reporting, and interpreting scientific data and apply this understanding in the context of physics investigations

Skill 3.1 Evaluating the appropriateness of a given method or procedure for collecting data for a specified purpose 23

Skill 3.2 Selecting an appropriate and effective graphic representation for organizing, reporting, and analyzing given experimental data 24

Skill 3.3 Applying procedures and criteria for formally reporting experimental procedures and data to the scientific community 26

Skill 3.4 Analyzing relationships between factors as indicated by experimental data 28

Skill 3.5 Coordinating explanations at different levels of scale, points of focus, and degrees of complexity and specificity 30

Skill 3.6 Applying statistical techniques to analyze data 31

Competency 4.0 Understand principles and procedures of measurement and the safe and proper use of equipment and materials used in physics investigations

Skill 4.1 Evaluating the appropriateness of SI units of measurement, measuring devices, or methods of measurement for given situations 33

Skill 4.2 Analyzing likely sources of error in given measurements in physics experiments 34

Skill 4.3 Distinguishing between accuracy and precision in scientific measurements 36

Skill 4.4 Using significant figures in performing calculations and expressing measurements 37

TEACHER CERTIFICATION STUDY GUIDE

Skill 4.5	Analyzing the principles upon which given laboratory instruments are based 38
Skill 4.6	Analyzing hazards of laboratory materials....................................40
Skill 4.7	Applying safety rules regarding electricity and electrical equipment 41
Skill 4.8	Applying proper procedures for dealing with accidents and injuries in the physics laboratory..43

Competency 5.0 Understand the use of mathematics and mathematical modeling in physics

Skill 5.1	Using mathematics to derive and solve equations 44
Skill 5.2	Applying algebra and geometry to model physical situations 45
Skill 5.3	Applying dimensional analysis to solve problems.........................47
Skill 5.4	Applying trigonometry and graphing to vectors 48
Skill 5.5	Using fundamental concepts of calculus 50

SUBAREA II   MECHANICS AND THERMODYNAMICS

Competency 6.0 Understand concepts related to motion in one and two dimensions, and apply this knowledge to solve problems that require the use of algebra, trigonometry, and graphing

Skill 6.1	Applying the terminology, units, and equations used to describe and analyze one and two-dimensional motion..............................52
Skill 6.2	Solving problems involving distance, displacement, speed, velocity, and constant acceleration..53
Skill 6.3	Interpreting information presented in one or more graphic representations related to distance, displacement, speed, velocity, and constant acceleration..56
Skill 6.4	Analyzing the movement of freely falling objects near the surface of the earth 57

PHYSICS

TEACHER CERTIFICATION STUDY GUIDE

Competency 7.0   **Understand characteristics of forces and methods used to measure force, and solve problems involving forces**

Skill 7.1   Identifying and analyzing the characteristics of the fundamental forces of nature 58

Skill 7.2   Identifying forces acting in a given situation 59

Skill 7.3   Analyzing experimental designs for measuring forces................ 60

Skill 7.4   Applying graphic solutions to solve problems involving the vector nature of force 62

Skill 7.5   Determining the resultant of two or more forces algebraically 63

Skill 7.6   Applying the concepts of force, pressure, and density 64

Skill 7.7   Applying Pascal's principle to analyze fluid statics 65

Skill 7.8   Applying Archimedes' principle to problems involving buoyancy and flotation 66

Competency 8.0   **Understand and apply the laws of motion and conservation of momentum**

Skill 8.1   Analyzing the characteristics of each of Newton's laws of motion and giving examples of each 68

Skill 8.2   Applying Newton's laws of motion in solving problems................ 70

Skill 8.3   Solving problems involving gravitational and frictional forces........ 71

Skill 8.4   Solving problems involving force constants................................ 74

Skill 8.5   Applying the conservation of momentum to analyze and solve problems 75

Skill 8.6   Understanding the implications of special relativity for the laws of motion 77

Competency 9.0   **Understand and apply the principle of conservation of energy and the concepts of energy, work, and power**

Skill 9.1   Analyzing mechanical systems in terms of work, power, and conservation of energy 79

PHYSICS

TEACHER CERTIFICATION STUDY GUIDE

Skill 9.2 Using the concept of conservation of energy to solve problems....82

Skill 9.3 Solving problems using the work-energy theorem..........................83

Skill 9.4 Determining power and efficiency as they relate to work and energy in a variety of situations 84

Skill 9.5 Analyzing energy conversions in everyday-life.............................84

Skill 9.6 Analyzing systems in which total energy is conserved but mechanical energy is not conserved 85

Competency 10.0 Understand the characteristics of circular motion, simple harmonic motion, and other periodic motion, and solve problems involving these types of motion

Skill 10.1 Applying vector analysis to describe uniform circular motion in radians..87

Skill 10.2 Determining the magnitude and direction of the force acting on a particle in uniform circular motion...89

Skill 10.3 Applying characteristics of circular motion and gravitational force to analyze planetary motion..90

Skill 10.4 Analyzing the relationships among displacement, velocity, and acceleration in simple harmonic motion..91

Skill 10.5 Analyzing energy interactions in oscillating systems 93

Skill 10.6 Solving equilibrium problems involving torques.............................94

Skill 10.7 Applying the relationship between torque and angular acceleration to solve problems 96

Skill 10.8 Applying the law of conservation of angular momentum to describe phenomena and solve problems 97

Skill 10.9 Applying the law of conservation of energy to systems involving rotational motion 99

Competency 11.0 Understand and apply the principles and laws of thermodynamics

Skill 11.1 Analyzing systems in terms of heat energy, internal energy, and work..100

PHYSICS

TEACHER CERTIFICATION STUDY GUIDE

Skill 11.2   Applying the first law of thermodynamics to analyze energy conversions in a variety of situations 100

Skill 11.3   Demonstrating an understanding of the second law of thermodynamics and of how the entropy of a system changes ... 102

Skill 11.4   Analyzing temperature and temperature scales 104

Skill 11.5   Solving problems involving thermal expansion and thermal contraction of solids, liquids, and ideal gases 105

Skill 11.6   Analyzing methods of heat transfers 108

Skill 11.7   Solving problems involving heat capacity, specific heat, heat transfer, heat of fusion, heat of vaporization, and phase changes 109

Skill 11.8   Describing thermal properties of matter in terms of molecular theory 113

SUBAREA III   ELECTRICITY AND MAGNETISM

Competency 12.0   Understand characteristics and units of electric charge, electric fields, electric potential, and capacitance; and apply principles of static electricity to problems involving Coulomb's law and electric field intensity

Skill 12.1   Analyzing the behavior of an electroscope 115

Skill 12.2   Analyzing common electrostatic phenomena 117

Skill 12.3   Applying Coulomb's law to determine the forces between charges 117

Skill 12.4   Applying principles of electrostatics to determine electric field intensity 119

Skill 12.5   Analyzing the vector nature of electric fields and forces 120

Skill 12.6   Applying principles of electrostatics to determine electrical potential 120

Skill 12.7   Applying the relationship among capacitance, charge, and potential difference 122

TEACHER CERTIFICATION STUDY GUIDE

Competency 13.0 Understand and analyze characteristics of DC circuits

Skill 13.1 Analyzing a DC circuit in terms of conservation of energy and conservation of charge 125

Skill 13.2 Analyzing the interrelationships among potential difference, resistance, and current 126

Skill 13.3 Analyzing factors that affect resistance 127

Skill 13.4 Solving problems involving Ohm's law 129

Skill 13.5 Describing how to use various meters to measure properties of an electric circuit 129

Skill 13.6 Interpreting schematic diagrams of electric circuits 130

Skill 13.7 Analyzing work, energy, and power in DC circuits 131

Skill 13.8 Describing the properties of conductors, semiconductors, superconductors 132

Skill 13.9 Describing the function of a solid state device in an electric circuit 134

Competency 14.0 Understand magnets, electromagnets, and magnetic fields; the effects of magnetic fields on moving electric charges; and the applications of electromagnetism

Skill 14.1 Describing magnetic fields as the result of moving electric charges 136

Skill 14.2 Applying the domain theory to the magnetization of ferromagnetic materials 138

Skill 14.3 Determining the orientation and magnitude of a magnetic field ... 139

Skill 14.4 Determining the magnitude and direction of the force on a charge or charges moving in a magnetic field 140

Skill 14.5 Analyzing the behavior of a current-carrying wire in a magnetic field 141

Skill 14.6 Analyzing factors that affect the strength of an electromagnet 142

Skill 14.7 Analyzing the use of electromagnetism in technology 143

PHYSICS

TEACHER CERTIFICATION STUDY GUIDE

Competency 15.0 Understand and apply the principles of electromagnetic induction and AC circuits

Skill 15.1   Analyzing factors that affect the magnitude of an induced electromotive force 145

Skill 15.2   Applying the appropriate hand rule to determine the direction of an induced current 146

Skill 15.3   Analyzing Lenz's law in terms of energy conservation 147

Skill 15.4   Analyzing transformers and generators 148

Skill 15.5   Analyzing an AC circuit, including relationships involving impedance, reactance, and resonance 150

Skill 15.6   Describing how AC and DC can be converted from one form to another 152

SUBAREA IV   WAVES, SOUND, AND LIGHT

Competency 16.0 Understand the properties and behavior of waves

Skill 16.1   Demonstrating knowledge of how waves transfer energy and momentum 154

Skill 16.2   Comparing types and characteristics of waves 155

Skill 16.3   Applying the wave properties to determine a wave's velocity, wavelength, or frequency 156

Skill 16.4   Analyzing the transmission and absorption of waves 157

Skill 16.5   Analyzing the reflection, refraction, dispersion, diffraction, and polarization of waves 159

Skill 16.6   Applying the superposition principle to determine characteristics of a resultant wave 161

Skill 16.7   Solving problems involving wave phenomena 162

Competency 17.0 Understand and apply knowledge of the characteristics, production, and transmission of sound waves

Skill 17.1   Analyzing the physical nature of sound waves, including intensity and frequency and how they relate to loudness and pitch, respectively 165

TEACHER CERTIFICATION STUDY GUIDE

Skill 17.2 Analyzing factors that affect the speed of sound in different media 166

Skill 17.3 Analyzing characteristics of standing waves in musical instruments 167

Skill 17.4 Analyzing situations involving resonance, harmonics, and overtones 168

Skill 17.5 Analyzing and solving problems involving the Doppler effect 170

Competency 18.0 Understand the production and characteristics of electromagnetic waves

Skill 18.1 Analyzing properties of the electromagnetic spectrum 172

Skill 18.2 Analyzing color when light is transmitted, absorbed, and reflected 172

Skill 18.3 Analyzing variations in energy, frequency, and amplitude in terms of the vibrations of the sources that produce them 174

Skill 18.4 Analyzing practical applications of the components of the electromagnetic spectrum 175

Skill 18.5 Describing how the constancy of the speed of light in a vacuum influences relationships among space, time, and energy 177

Skill 18.6 Analyzing and solving problems involving the Doppler effect 179

Competency 19.0 Understand and apply the characteristics of reflective and refractive devices

Skill 19.1 Comparing types and characteristics of lenses and mirrors 180

Skill 19.2 Using a ray diagram to locate the focal point or point of image formation of a lens or mirror 182

Skill 19.3 Applying the lens and mirror equations to analyze problems involving lenses and mirrors 185

Skill 19.4 Applying Snell's law to analyze optical phenomena 187

Skill 19.5 Analyzing applications of lenses, mirrors, and prisms 188

PHYSICS

TEACHER CERTIFICATION STUDY GUIDE

| SUBAREA V | QUANTUM THEORY AND THE ATOM |

Competency 20.0 Understand the dual nature of light and matter

Skill 20.1   Describing the quantization of energy in terms of Planck's theory 191

Skill 20.2   Applying the laws of photoelectric emission to explain photoelectric phenomena 192

Skill 20.3   Applying the principles of stimulated emission of radiation to lasers and masers 193

Skill 20.4   Analyzing evidence supporting the dual nature of light and matter 195

Skill 20.5   Solving problems using de Broglie's equation 196

Skill 20.6   Applying the uncertainty principle 197

Competency 21.0 Understand physical models of atomic structure and the nature of elementary particles

Skill 21.1   Analyzing historic and contemporary models of atomic structure 199

Skill 21.2   Interpreting notation used to represent elements, molecules, ions, and isotopes 200

Skill 21.2   Analyzing bright-line spectra in terms of electron transitions 201

Skill 21.3   Recognizing the relationship between the design of particle accelerators and elementary particle characteristics 204

Competency 22.0 Understand the standard model of particle physics

Skill 22.1   Recognizing key contributions by individuals in the development of the standard model 206

Skill 22.2   Describing historical developments leading to the standard model 207

Skill 22.3   Describing experimental techniques and methods used to investigate elementary particles 209

Skill 22.4   Describing the families of subnuclear particles and their properties 210

Skill 22.5	Explaining the properties of quarks and how they interact to form protons, neutrons, and other particles 211
Skill 22.6	Describing relationships between particles and antiparticles....... 211

Competency 23.0 Understand the principles of radioactivity and nuclear reactions and their applications

Skill 23.1	Applying principles of the conservation of mass number and charge to balance equations for nuclear reactions 213
Skill 23.2	Analyzing radioactive decay including the half-life concept and the nuclear disintegration series for a given isotope......................... 215
Skill 23.3	Describing the basic operation of types of radiation detectors 217
Skill 23.4	Analyzing characteristics of fission and components of a nuclear reactor, including the problems associated with operating a nuclear reactor 218
Skill 23.5	Analyzing fusion reactions and their application to the sun's energy and astronomy.. 220
Skill 23.6	Applying the principle of conservation of mass-energy to calculate nuclear mass defect and binding energy 221

Sample Test 225

Free Response Question... 245

Answer Key 247

Rigor Analysis Table 248

Rationales with Sample Questions 249

TEACHER CERTIFICATION STUDY GUIDE

Great Study and Testing Tips!

What to study in order to prepare for the subject assessments is the focus of this study guide but equally important is *how* you study.

You can increase your chances of truly mastering the information by taking some simple, but effective steps.

Study Tips:

1. Some foods aid the learning process. Foods such as milk, nuts, seeds, rice, and oats help your study efforts by releasing natural memory enhancers called CCKs (*cholecystokinin*) composed of *tryptophan*, *choline*, and *phenylalanine*. All of these chemicals enhance the neurotransmitters associated with memory. Before studying, try a light, protein-rich meal of eggs, turkey, and fish. All of these foods release the memory enhancing chemicals. The better the connections, the more you comprehend.

Likewise, before you take a test, stick to a light snack of energy boosting and relaxing foods. A glass of milk, a piece of fruit, or some peanuts all release various memory-boosting chemicals and help you to relax and focus on the subject at hand.

2. Learn to take great notes. A by-product of our modern culture is that we have grown accustomed to getting our information in short doses (i.e. TV news sound bites or USA Today style newspaper articles.)

Consequently, we've subconsciously trained ourselves to assimilate information better in neat little packages. If your notes are scrawled all over the paper, it fragments the flow of the information. Strive for clarity. Newspapers use a standard format to achieve clarity. Your notes can be much clearer through use of proper formatting. A very effective format is called the *"Cornell Method."*

> Take a sheet of loose-leaf lined notebook paper and draw a line all the way down the paper about 1-2" from the left-hand edge.
>
> Draw another line across the width of the paper about 1-2" up from the bottom. Repeat this process on the reverse side of the page.

Look at the highly effective result. You have ample room for notes, a left hand margin for special emphasis items or inserting supplementary data from the textbook, a large area at the bottom for a brief summary, and a little rectangular space for just about anything you want.

PHYSICS

TEACHER CERTIFICATION STUDY GUIDE

3. <u>Get the concept then the details</u>. Too often we focus on the details and don't gather an understanding of the concept. However, if you simply memorize only dates, places, or names, you may well miss the whole point of the subject.

A key way to understand things is to put them in your own words. If you are working from a textbook, automatically summarize each paragraph in your mind. If you are outlining text, don't simply copy the author's words.

Rephrase them in your own words. You remember your own thoughts and words much better than someone else's, and subconsciously tend to associate the important details to the core concepts.

4. <u>Ask Why?</u> Pull apart written material paragraph by paragraph and don't forget the captions under the illustrations.

Example: If the heading is "Stream Erosion", flip it around to read "Why do streams erode?" Then answer the questions.

If you train your mind to think in a series of questions and answers, not only will you learn more, but it also helps to lessen the test anxiety because you are used to answering questions.

5. <u>Read for reinforcement and future needs</u>. Even if you only have 10 minutes, put your notes or a book in your hand. Your mind is similar to a computer; you have to input data in order to have it processed. *By reading, you are creating the neural connections for future retrieval.* The more times you read something, the more you reinforce the learning of ideas.

Even if you don't fully understand something on the first pass, *your mind stores much of the material for later recall.*

6. <u>Relax to learn so go into exile</u>. Our bodies respond to an inner clock called biorhythms. Burning the midnight oil works well for some people, but not everyone.

If possible, set aside a particular place to study that is free of distractions. Shut off the television, cell phone, and pager and exile your friends and family during your study period.

If you really are bothered by silence, try background music. Light classical music at a low volume has been shown to aid in concentration over other types. Music that evokes pleasant emotions without lyrics is highly suggested. Try just about anything by Mozart. It relaxes you.

PHYSICS

7. Use arrows not highlighters. At best, it's difficult to read a page full of yellow, pink, blue, and green streaks. Try staring at a neon sign for a while and you'll soon see that the horde of colors obscure the message.

A quick note, a brief dash of color, an underline, and an arrow pointing to a particular passage is much clearer than a horde of highlighted words.

8. Budget your study time. Although you shouldn't ignore any of the material, *allocate your available study time in the same ratio that topics may appear on the test.*

TEACHER CERTIFICATION STUDY GUIDE

Testing Tips:

1. Get smart, play dumb. Don't read anything into the question. Don't make an assumption that the test writer is looking for something else than what is asked. Stick to the question as written and don't read extra things into it.

2. Read the question and all the choices *twice* before answering the question. You may miss something by not carefully reading, and then re-reading both the question and the answers.

If you really don't have a clue as to the right answer, leave it blank on the first time through. Go on to the other questions, as they may provide a clue as to how to answer the skipped questions.

If later on, you still can't answer the skipped ones . . . *Guess.* The only penalty for guessing is that you *might* get it wrong. Only one thing is certain; if you don't put anything down, you will get it wrong!

3. Turn the question into a statement. Look at the way the questions are worded. The syntax of the question usually provides a clue. Does it seem more familiar as a statement rather than as a question? Does it sound strange?

By turning a question into a statement, you may be able to spot if an answer sounds right, and it may also trigger memories of material you have read.

4. Look for hidden clues. It's actually very difficult to compose multiple-foil (choice) questions without giving away part of the answer in the options presented. In most multiple-choice questions you can often readily eliminate one or two of the potential answers. This leaves you with only two real possibilities and automatically your odds go to Fifty-Fifty for very little work.

5. Trust your instincts. For every fact that you have read, you subconsciously retain something of that knowledge. On questions that you aren't really certain about, go with your basic instincts. **Your first impression on how to answer a question is usually correct.**

6. Mark your answers directly on the test booklet. Don't bother trying to fill in the optical scan sheet on the first pass through the test.

Just be very careful not to miss-mark your answers when you eventually transcribe them to the scan sheet.

7. Watch the clock! You have a set amount of time to answer the questions. Don't get bogged down trying to answer a single question at the expense of 10 questions you can more readily answer.

PHYSICS

TEACHER CERTIFICATION STUDY GUIDE

SUBAREA I	FOUNDATIONS OF SCIENTIFIC INQUIRY

COMPETENCY 1.0 UNDERSTANDS THE HISTORICAL AND CONTEMPORARY CONTEXTS OF THE STUDY OF PHYSICS, INCLUDING THE RELATIONSHIPS AND COMMON THEMES THAT CONNECT MATHEMATICS, SCIENCE, AND TECHNOLOGY AND THEIR APPLICATIONS TO EVERYDAY LIFE

Skill 1.1 Analyzing the role of early civilizations in establishing the foundations of physics (e.g., the law of the lever, hydrostatics, the idea of the atom)

Though people have tried to explain the phenomena of the natural world for millions of year, physics as we know it largely begins with the Greek scholars and continues with the Romans. Scientists in these civilizations were interested in the nature of matter, plant and animal life including the human body, simple mechanics, and the movement of heavenly bodies. Their scientific progress was somewhat impeded by the lack of accurate measuring devices and other advanced experimental equipment. Additionally, the loss of many written records (most notably as a result of the destruction of the Library of Alexandria) means that we do not have a full picture of their scientific knowledge.

It is known, however, that the Greeks did make many discoveries and formulate hypotheses that are still regarded as correct today. For instance, Greek mathematicians devised methods to calculate the area of three-dimensional objects that foreshadow integral calculus. Early Greek chemists speculated on the concept of a very tiny, indivisible form of matter (the atom) although they did not understand that each element was composed from a unique type of atom. Archimedes derived many correct principles of mechanics and hydrostatics (including his legendary "Eureka" moment that occurred when he noticed his own body displacing water in a bath). He also devised the Archimedes screw to transport water upward from rivers to farmland and formulated the law of the lever. The Greeks also possessed a surprising amount of knowledge about astronomy. Eratosthenes, for example, was able to determine that the Earth was a sphere and very accurately compute its circumference. Additionally, Aristarchus of Samos deduced that the Earth rotated on an axis and orbited around the sun yearly. Scientific progress continued within the Roman Empire but their most important contributions to physics are related to their feats of engineering (roads, bridges, aqueducts, etc).

In the Western world, we tend to be most familiar with the early science and engineering of the Greeks and Romans. Unfortunately, the lack of written records often means that we know little about the science and technology of other ancient people, for instance, the native people of the Americas. However, there is a great deal of evidence that many other early civilizations were quite scientifically advanced. For instance, ancient Indian astronomers made many accurate calculations of the Earth's size and had a heliocentric model of the solar system. They also had a fairly advanced understanding of chemistry and chemical reactions, including a rudimentary notion of the atom. Indian mathematicians were particularly advanced and possessed knowledge of trigonometric functions and the concept of zero. Chinese scientists were similarly advanced in the fields of mathematics and astronomy. The Chinese were particularly gifted inventors and engineers. They were early developers of many technologies including abacuses, sundials, gunpowder, compasses, wheelbarrows, suspension bridges, parachutes, propellers, crossbows, and processes for papermaking and printing.

Unlike the Indian and Chinese scientists, Muslim scholars had a relatively high amount of interaction with their counterparts in the Western World. Geography and cultural forces led to much cross-over with their early Greek and Roman counterparts. Persian scientists invented both algebra and the astrolabe which was used by scientists in the East and West to advance the study of the heavenly bodies. The Iraqi physicist Alhazen studied and then improved upon Ptolemy's theory of light refraction. His early discoveries in the field of optics are still used today. Perhaps the greatest contribution of early Islamic scientists was their insistence on experimentation which was instrumental in developing the modern scientific method.

Skill 1.2 Recognizing key events and analyzing the contributions of individuals in the development of mechanics, electromagnetism, the kinetic theory, atomic theory, radioactivity, quantum theory, and solid state physics

Archimedes
Archimedes was a Greek mathematician, physicist, engineer, astronomer, and philosopher. He is credited with many inventions and discoveries some of which are still in use today such as the Archimedes screw. He designed the compound pulley, a system of pulleys used to lift heavy loads such as ships.

Although Archimedes did not invent the lever, he gave the first rigorous explanation of the principles involved which are the transmission of force through a fulcrum and moving the effort applied through a greater distance than the object to be moved. His Law of the Lever states that magnitudes are in equilibrium at distances reciprocally proportional to their weights.

He also laid down the laws of flotation and described Archimedes' principle which states that a body immersed in a fluid experiences a buoyant force equal to the weight of the displaced fluid.

Niels Bohr

Bohr was a Danish physicist who made fundamental contributions to understanding atomic structure and quantum mechanics. Bohr is widely considered one of the greatest physicists of the twentieth century.
Bohr's model of the atom was the first to place electrons in discrete quantized orbits around the nucleus.

Bohr also helped determine that the chemical properties of an element are largely determined by the number of electrons in the outer orbits of the atom. The idea that an electron could drop from a higher-energy orbit to a lower one emitting a photon of discrete energy originated with Bohr and became the basis for future quantum theory.

He also contributed significantly to the Copenhagen interpretation of quantum mechanics. He received the Nobel Prize for Physics for this work in 1922.

Marie Curie

Curie was as a Polish-French physicist and chemist. She was a pioneer in radioactivity and the winner of two Nobel Prizes, one in Physics and the other in Chemistry. She was also the first woman to win the Nobel Prize.

Curie studied radioactive materials, particularly pitchblende, the ore from which uranium was extracted. The ore was more radioactive than the uranium extracted from it which led the Curies (Marie and her husband Pierre) to discover a substance far more radioactive then uranium. Over several years of laboratory work the Curies eventually isolated and identified two new radioactive chemical elements, polonium and radium. Curie refined the radium isolation process and continued intensive study of the nature of radioactivity.

Albert Einstein

Einstein was a German-born theoretical physicist who is widely considered one of the greatest physicists of all time. While best known for the theory of relativity, and specifically mass-energy equivalence, $E = mc^2$, he was awarded the 1921 Nobel Prize in Physics for his explanation of the photoelectric effect and "for his services to Theoretical Physics". In his paper on the photoelectric effect, Einstein extended Planck's hypothesis ($E = h\nu$) of discrete energy elements to his own hypothesis that electromagnetic energy is absorbed or emitted by matter in quanta and proposed a new law $E_{max} = h\nu - P$ to account for the photoelectric effect.

He was known for many scientific investigations including the special theory of relativity which stemmed from an attempt to reconcile the laws of mechanics with the laws of the electromagnetic field. His general theory of relativity considered all observers to be equivalent, not only those moving at a uniform speed. In general relativity, gravity is no longer a force, as it is in Newton's law of gravity, but is a consequence of the curvature of space-time.

Other areas of physics in which Einstein made significant contributions, achievements or breakthroughs include relativistic cosmology, capillary action, critical opalescence, classical problems of statistical mechanics and problems in which they were merged with quantum theory (leading to an explanation of the Brownian movement of molecules), atomic transition probabilities, the quantum theory of a monatomic gas, the concept of the photon, the theory of radiation (including stimulated emission), and the geometrization of physics.

Einstein's research efforts after developing the theory of general relativity consisted primarily of attempts to generalize his theory of gravitation in order to unify and simplify the fundamental laws of physics, particularly gravitation and electromagnetism, which he referred to as the Unified Field Theory.

Michael Faraday
Faraday was an English chemist and physicist who contributed significantly to the fields of electromagnetism and electrochemistry. He established that magnetism could affect rays of light and that the two phenomena were linked. It was largely due to his efforts that electricity became viable for use in technology. The unit for capacitance, the farad, is named after him as is the Faraday constant, the charge on a mole of electrons (about 96,485 coulombs). Faraday's law of induction states that a magnetic field changing in time creates a proportional electromotive force.

J. Robert Oppenheimer
Oppenheimer was an American physicist best known for his role as the scientific director of the Manhattan Project, the effort to develop the first nuclear weapons. Sometimes called "the father of the atomic bomb", Oppenheimer later lamented the use of atomic weapons. He became a chief advisor to the United States Atomic Energy Commission and lobbied for international control of atomic energy. Oppenheimer was one of the founders of the American school of theoretical physics at the University of California, Berkeley. He did important research in theoretical astrophysics, nuclear physics, spectroscopy, and quantum field theory.

Sir Isaac Newton

Newton was an English physicist, mathematician, astronomer, alchemist, and natural philosopher in the late 17th and early 18th centuries. He described universal gravitation and the three laws of motion laying the groundwork for classical mechanics. He was the first to show that the motion of objects on earth and in space is governed by the same set of mechanical laws. These laws became central to the scientific revolution that took place during this period of history. Newton's three laws of motion are:

> I. Every object in a state of uniform motion tends to remain in that state of motion unless an external force is applied to it.
> II. The relationship between an object's mass m, its acceleration a, and the applied force F is $F = ma$.
> III. For every action there is an equal and opposite reaction.

In mechanics, Newton developed the basic principles of conservation of momentum. In optics, he invented the reflecting telescope and discovered that the spectrum of colors seen when white light passes through a prism is inherent in the white light and not added by the prism as previous scientists had claimed. Newton notably argued that light is composed of particles. He also formulated an experimental law of cooling, studied the speed of sound, and proposed a theory of the origin of stars.

Erwin Schrödinger

Schrödinger was a Nobel Prize winning Austrian physicist who is best remembered for his contributions to quantum mechanics. Chief among his findings was the Schrödinger equation which describes the space and time-dependence of quantum systems. This equation predicts behavior of microscopic particles in much the same way that Newton's second law predicts the behavior of macroscopic matter. Much of solid state physics was built upon the Schrödinger (wave) equation. Schrödinger also devised the "cat in a box" thought experiment to illustrate a paradox of quantum mechanics.

Enrico Fermi

Fermi was an extremely gifted Italian physicist and Nobel Prize winner. His work focused on radioactivity, quantum theory, and statistical mechanics and he helped develop the first nuclear reactor. He also participated in the Manhattan Project after fleeing to America prior to the Second World War. Fermi's Golden Rule is an important equation in quantum mechanics and is used to calculate the transition rates of quantum systems. Much of Fermi's work in quantum mechanics (Fermi holes, Fermi levels, Fermi-Dirac statistics) ultimately led to our modern understanding of semiconductors.

James Clerk Maxwell

Maxwell was a Scottish theoretical physicist and mathematician whose work was especially important to our understanding of electromagnetism. His famous Maxwell's equations were the first to unify and simply express the laws of electricity and magnetism. Maxwell was also important in the development of the kinetic theory; the Maxwell-Boltzman distribution is a probability distribution used in statistical mechanics and to predict the behavior of gases. His work paved the way for much of the progress in physics during the 20th century (special relativity, quantum mechanics, etc) and ultimately led to all modern electrical and communications technology.

Daniel Bernoulli

Bernoulli was a Swiss mathematician who developed many equations important in physics. His work was especially focused on the behavior of fluids and the Bernoulli equation governs all steady, inviscid, incompressible flow. The resultant Bernoulli principle is especially useful in aerodynamics. The Euler-Bernoulli beam theory, which allows calculation of the loading characteristics of a beam, was of key importance during the Industrial Revolution and remains an important equation for engineers today. Bernoulli was also the first scientist to suggest the principles of the kinetic theory of gases.

Skill 1.3   Demonstrating an understanding of current theories of the origin, structure, and composition of the universe

Cosmologists typically use the term universe to refer to the **observable universe**. That is the finite part of space-time that we can study using various instruments and theoretical or empirical physics. However, this observable universe is thought to be a part of a much vaster theoretical universe that cannot be observed. It is not presently understood whether the observable and theoretical universe are nearly the same size or whether the theoretical universe is infinitely large. Below, we confine our discussion to the observable universe.

For millions of years, people have speculated about the origin of our world and universe. The **Big Bang theory** is currently the most widely accepted scientific explanation. This theory was originally devised after it was observed that the universe is presently expanding. This expansion implies that in the universe's distant past, there must have been a gravitational singularity at which all the matter in the universe was compressed into an infinitesimal point.
Approximately 13.7 billion years ago, this point began to expand and a universe of hot, dense plasma (quark-gluon plasma) was created. As the universe continued to expand, the temperature dropped to the point at which matter and anti-matter could begin to form. Atoms began to form and energy, previously coupled to matter, was able to travel through space. Eventually, in denser regions, some matter became gravitationally attracted to other matter and bodies including stars, planets, and entire galaxies were formed.

Investigations continue into how the universe has evolved following the Big Bang and what it looks like today. Specifically, the size and shape of the universe are unknown. **The size of the visible universe has been suggested to be anywhere from 13.7 to 180 billion light years.** Most cosmologists believe that the universe is nearly flat but that it has "wrinkles" in areas where large objects distort space-time. Also under investigation is whether or not the universe is spatially finite (i.e., multiply or singly connected).

Best estimates are that the universe consists of 73% dark energy, 23% cold dark matter and 4% atoms. Dark energy is a hypothetical type of energy that can permeate all space and increases the universe's rate of expansion. Cold dark matter is composed of very slowly moving particles and cannot be detected by observing its electromagnetic radiation. Neither dark energy nor cold dark matter is well understood currently.

Skill 1.4 Assessing the societal implications of developments in physics (e.g., heliocentric theory, nuclear technology, the transistor, the laser)

Nearly all advances in science and technology have some effect on society. However, some are especially important in changing the way we think about the natural world or how we lead our day to day lives. Several examples drawn from physics are briefly explored below.

Heliocentric theory
While we largely take for granted that the planets in our solar system orbit around the sun, during the 16th and 17th centuries it was a controversial notion in the western world. Most religious authorities, at least publicly, condemned the writings of Galileo and Copernicus which postulated a heliocentric model of the solar system. Both scientists and theologians argued over how new scientific evidence could be brought into agreement with scripture. This provides a famous historical example of how organized religion can interact with, and at times impede, scientific progress. Today, we still see this playing out in debates over such theories as the Big Bang and evolution.

Nuclear physics and the atomic bomb

Advances in nuclear physics in the first half of the 20th century led to the realization that tremendous amounts of power could be generated for fuel or weaponry through atomic fission and fusion. During World War II, the United State and its allies (through the Manhattan Project) were the first to succeed in the creation of atomic bombs which were subsequently dropped on Hiroshima and Nagasaki in 1945. The long term and exceedingly devastating effect of these weapons cannot be overstated. Nuclear proliferation (especially by the US and the USSR) in the following 50 years contributed to the Cold War and created an imminent threat of vast worldwide destruction. While the Cold War has since ended, the continued testing and possession of nuclear weaponry by an ever growing number of countries remains an important socio-political problem.

Nuclear power is currently use but remains controversial. On one hand, the extreme short and long term dangers of nuclear accidents have been demonstrated by incidents such as those at Three Mile Island and Chernobyl. On the other hand, society is highly in need of a cleaner, more sustainable source of energy and nuclear power might be able to fill this role.

Semiconductors, the transistor, and digital devices

The discovery that the electrical conductive properties of semiconductors can be either permanently or temporarily manipulated has been an important one. Most importantly, it led to the creation of solid state electrical devices such as transistors. In electrical circuits, these elements replaced the previously used vacuum tubes and mechanical relays which were highly susceptible to wear and physical stressors. Transistors, however, are small, fast, accurate, reliable, and can be produced quickly and cheaply. Thus, transistors have paved for the way for the vast array of affordable electrical devices. They have been particularly important in advancing computers and digital technology. We are well aware of how digital devices and information technology, made possible in part by transistors, touch every part of our modern lives.

MEMS and nanotechnology

Many people speculate that nanotechnology is poised to make significant changes in our lives. Microelectromechanical Systems (MEMS), or micromachines are made possible by modern materials and technologies (such as lasers and micro-etching) that allow manufacture and manipulation of devices on the micrometer scale. Developments in these technologies are already being used in the automotive, biomedical, chemical, cosmetic, telecommunications, and manufacturing industries.

TEACHER CERTIFICATION STUDY GUIDE

Skill 1.5 **Applying the laws of physics to geological, chemical, biological, and astronomical systems**

The laws of physics are those that govern all parts of the natural world. The laws of physics reveal the root explanation for every geological, chemical, biological, and astronomical phenomenon. Let's examine how knowledge of the different aspects of physics is useful in understanding phenomenon from each field.

Geology
The study of the physical laws that affect the earth and its processes is known as geophysics. Thermodynamics is needed to understand how heat moves between the core, mantle, and crust of the earth. Additionally, thermodynamics, along with the study of electromagnetism, is needed to comprehend and predict meteorological events. Mechanics is required to analyze the movement of the tectonic plates, earthquakes, and other seismic activity. Mechanics, particularly those of waves, is also key in the study of hydrology.

Chemistry
Physical chemistry is the branch of chemistry concerned with physical behavior of molecules and the physical aspects of chemical reactions. Concepts from modern physics including nuclear physics and quantum and statistical mechanics are of key importance in this field. Thermodynamics and basic mechanics are also useful in understanding the progress of chemical reactions. Chemists have also made extensive use of the laws of electromagnetism in designing analysis tools such as NMR spectrometers.

Biology
The application of physical theories and methods in biological systems is known as biophysics. This field is largely devoted to the micromechanics of the cell, particularly the cytoskeleton and the physical structure of proteins. However, understanding larger biological systems also requires an understanding of physics. For instance, thermodynamics is needed to analyze heat transfer in warm and cold blooded animals. Mechanics is necessary to understand and develop therapies for the musculoskeletal system.

Astronomy
Astrophysics is the study of the physics of the universe. Mechanics is important in predicting the motion of planets, comets, and other celestial bodies. Thermodynamics is used in studying the transfer of energy and the heating and cooling of planets and stars. Optics and an understanding of sound and light are also necessary, particularly for the proper use of telescopes and other instruments used for studying the universe.

Skill 1.6 **Analyzing the use of physics, mathematics, and other sciences in the design of a technological solution to a given problem**

Technology includes all types of devices and methods created by humans to achieve a given objective. Simple tools, complex machines, manufacturing processes, and software are all types of technology. It is true that scientific reasoning and a grasp of scientific laws are not essential in the development of new technology. Indeed, our early ancestors devised many primitive technologies without the aid of science as we understand it. Various more modern inventions have also been created with little or no scientific assistance. However, as inventions become more complex, a firm grasp of scientific principles is needed to develop effective solutions for problems. Additionally, the development of new technology can be rapidly accelerated by taking advantages of scientific concepts.

Firstly, **science can help us in identifying problems and suggesting solutions**. For example, molecular biology and biochemistry are useful in helping us identify underlying problems in clinical conditions and discovering potential targets for pharmaceuticals. Similarly, various atmospheric and chemical measurements have been used to quantify climate change and suggest mechanisms. Additionally, a study of organisms can show us how evolution has solved problems. One classic example of this is the invention of Velcro which mimics the action of burrs on animal fur.

Second, **an understanding of a broad range of sciences may be helpful in predicting the suitability of various options and in the creation of new ones**. Consider, for instance, an implantable orthopedic device such as a hip replacement that must both support weight and be non-toxic. We can use mechanics to predict the forces that the device must withstand. Similarly, principles of material science help us determine what type of substance would have appropriate mechanical and chemical properties. If current substances do not meet the requirements, organic chemistry may allow the creation of new ones.

Finally, **scientific reasoning can assist in the process of testing and refining a new technical design**. The basics of the scientific method allow us to determine what tests should be done and what we can learn from each. Mathematics allows us to model the behavior of the technology. Statistics is also particularly useful in helping us determine how confident we can be in the results of the testing and whether additional experimentation is necessary. It may also help us quickly hone in on optimal conditions, for instance, in a new manufacturing process with multiple variables.

Skill 1.7 Analyzing the role of technology in the advancement of scientific knowledge

Scientific discoveries are important in the development of new technology. However, new technologies may also allow for further advances in scientific understanding. This is largely because new technologies allow us to observe the natural world in ways not previously possible. Additionally, new technology may allow us to store and analyze data in new ways. Finally, technological advances pique the interest of funding bodies and tax-payers thereby fueling further scientific investigations. Below are some examples of this phenomenon within the study of physics.

Astrophysics-space exploration
Though telescopes have been used by astronomers for hundreds of years, they have become increasingly powerful. With each improvement, scientists are able to obtain more detail about the behavior and composition of distant bodies. Dramatic advances in our knowledge of the galaxy were made during the 20th century's "space race." New developments in communications and materials allowed the creation of satellites, space probes, and manned missions that provided never before seen images of the earth and other planets and allowed new experimentation. The space race also provides us with an excellent example of how the public's interest in technology can help stimulate further scientific advances.

Biophysics-micro- and nano-instrumentation
At the other end of the size scale from telescopes, microscopes have also become more powerful over time. These have allowed us to visualize the make-up of sub-cellular structures. Even more interestingly, new tools for micro-manipulation (for example, laser tweezers) allow scientists to observe, modify and measure the behavior of proteins and nucleic acids.

Computational capabilities
Advances in computational ability have made possible the storage and manipulation of large amounts of data. These systems are then able to analyze the data for trends and correlations more rapidly and accurately than humans could. The result is highly accurate models of natural systems. Because these models are extremely complex with many variables, it is only possible to produce or use them with the assistance of powerful computers. Not only can these models be used to explain the effects of various factors, they can be used to perform simulations. In many cases, simulations allow us to investigate phenomena in which experiments are not feasible. Here again we see how the demand for new technologies can propel basic scientific research: the desire for smaller, faster electronic devices has stimulated a large amount of research into semiconductors.

Skill 1.8 **Evaluating the appropriateness of a variety of software (e.g., spreadsheets, graphing utilities, statistical packages, simulations) and information technologies to model and solve problems in mathematics, science, and technology**

Advances in information technology, including software, have been immensely helpful in all the physical sciences. Information technology has allowed for the rapid communication of experimental results and hypotheses between scientists and engineers working in distant locations. For example, the ability to communicate rapidly by email and send digital images and other files means scientists can easily confer on new data and interpretations. They can easily find and consult with experts outside their own field who might have particular knowledge. Information technology has also made it easier to find already published experimental results; many journals have online editions and search engines make finding information on a given topic straightforward.

Specific software packages can also assist the advancement of science. Some important types of software and their appropriate uses are detailed below:

Spreadsheets
Spreadsheets contain multiple pages of grids into which numbers and formulas can be entered. They are excellent tools for storing data and doing some simple mathematical manipulations. Many spreadsheets now include graphical and statistical packages (see below), though often they are not as powerful as "stand alone" software for those uses. Spreadsheets are not typically used for complex statistical analysis or simulations though they may have some rudimentary capabilities. The data would be imported it to another, more specific program if further analysis were needed.

Databases
Like spreadsheets, databases are designed for the storage and manipulation of data. However, they are much more sophisticated and have increased security and accountability control. Though they may sometimes be useful for managing extremely large or shared datasets, databases are not typically useful to scientists.

Graphing packages
Graphing packages allow data to be transformed into a variety of visual representations. Because graphing is so important for recognizing trends and communicating experimental results, scientists make extensive use of this type of software. Most graphical packages will also perform basic modeling or statistical analysis (for instance, fitting a line or determining standard deviation). This is appropriate in some simple cases but, most often, statistical packages are better suited for this type of analysis. Graphing software should usually be limited to displaying the results of this analysis (for example, with error bars).

TEACHER CERTIFICATION STUDY GUIDE

Statistical packages
Statistical packages perform all types of statistical analysis to determine if significant differences exist between datasets. Thus, this type of software is extremely helpful to scientists especially when one is trying to analyze trends or determine the effects of a given factor in a large, complex experiment. Though this software can simplify the performance of a statistical test, the scientist must still carefully determine *which* statistical test will provide her with the answer she seeks. Though statistical software packages may have tools for data storage and graphing, it is usually preferable to use these only temporarily or for a "first-pass" and ultimately use a more specific software package for these tasks.

Simulators
Simulators assist with the creation of complex mathematical models of the natural world. They also allow for simulated experiments in which the value of variables in the model are changed and hypothetical outcomes obtained. These simulators can be a tremendous boon to scientists when performing actual experiments would be expensive, unethical, or impossible. Though much important information can be obtained from these simulators, it is important that their findings be supported by experimental evidence whenever possible.

Skill 1.9   Analyzing solutions to everyday-life problems that incorporate knowledge and skills of mathematics, science, and technology

Physics can help us solve many everyday problems. Whether a child is trying to move a heavy object or a research engineer is designing a new medical diagnostic tool, knowledge in mathematics and science, particularly physics, is key. This could include an understanding of both basic physical laws and a comprehension of the technological developments that have resulted from physics research. Below are several broad areas in which physics has been useful in solving problems and improving life.

Simple machines
Familiar simple machines include pulleys, levers, screws, wheels with axles, and wedges. They all rely on basic mechanics to minimize the amount of force required to perform a task. Simple machines were some of the earliest technologies used to aid people in performing tasks.

Transportation and construction machinery

Often building on the basic mechanical principles seen in simple machines, more complex apparatus has been engineered for transportation and construction as well as for manufacturing. This category includes everything from bicycles to the internal combustion engine. Like simple machines, these technologies were used to help people perform their work more quickly and efficiently. Increased mechanization (among other factors) during the Industrial Revolution greatly changed how modern people live and work.

Communication and electrical devices

Our modern communication and information technologies largely rely on discoveries in the field of electromagnetism which paved the way for the first radios, telegraphs, and telephones. Since then, solid state physics research has allowed ever smaller and cheaper computing devices. Just as the Industrial Revolution exploited mechanization to change how people live and work, the digital revolution is using electronic devices to change it again. Reliable and easy data transfer, storage, and communication have helped create a mobile global economy.

Medical technology

A variety of concepts from physics have contributed to improvement in people's health. Biomechanics has assisted in the development of prosthetics and refined the practice of physical therapy. Similarly, an understanding of how repeated forces affect the body has created the field of ergonomics. Concepts from electromagnetism have been key in developing diagnostic imaging tools such as Doppler, ultrasound, and MRIs.

Competency 2.0 Understand the process of scientific inquiry and the role of observation and experimentation in explaining natural phenomena

Skill 2.1 Analyzing processes by which new scientific knowledge and hypotheses are generated

The process of science always begins with a question about how the natural world works. Typically, this occurs when a scientists makes an observation and asks why something is happening, what factors affect it, or how the outcome might be different in a slightly different situation. The scientist then formulates a possible explanation for what she has observed and develops experiments to determine if her explanation is correct.

This process has been codified into a system known as the scientific method. Though application of the scientific method may vary somewhat among different scientists and in different fields, the basic steps are diagrammed below:

Observations
Scientific questions result from observations of events in nature or events observed in the laboratory. An observation is not just a look at what happens. It also includes measurements and careful records of the event. Records could include photos, drawings, or written descriptions. The observations and data collection lead to a question. In physics, observations almost always deal with the behavior of matter. Having arrived at a question, a scientist usually researches the scientific literature to see what is known about the question. Maybe the question has already been answered. The scientist then may want to test the answer found in the literature. Or, maybe the research will lead to a new question.

Steps of a Scientific Method

Hypothesis
If the question has not been answered, the scientist may prepare for an experiment by making a hypothesis. A hypothesis is a statement of a possible answer to the question. It is a tentative explanation for a set of facts and can be tested by experiments. Although hypotheses are usually based on observations, they may also be based on a sudden idea or intuition.

Experiment
An experiment tests the hypothesis to determine whether it may be a correct answer to the question or a solution to the problem. Some experiments may test the effect of one thing on another under controlled conditions. Such experiments have two variables. The experimenter controls one variable, called the *independent variable*. The other variable, the *dependent variable*, is the change caused by changing the independent variable.

For example, suppose a researcher wanted to test the effect of vitamin A on the ability of rats to see in dim light. The independent variable would be the dose of Vitamin A added to the rats' diet. The dependent variable would be the intensity of light that causes the rats to react. All other factors, such as time, temperature, age, water given to the rats, the other nutrients given to the rats, and similar factors, are held constant.
Scientists sometimes do short experiments "just to see what happens". Often, these are not formal experiments. Rather they are ways of making additional observations about the behavior of matter.

In most experiments, scientists collect quantitative data, which is data that can be measured with instruments. They also collect qualitative data, descriptive information from observations other than measurements. Interpreting data and analyzing observations are important. If data is not organized in a logical manner, wrong conclusions can be drawn. Also, other scientists may not be able to follow your work or repeat your results.

Conclusion
Finally, a scientist must draw conclusions from the experiment. A conclusion must address the hypothesis on which the experiment was based. The conclusion states whether or not the data supports the hypothesis. If it does not, the conclusion should state what the experiment *did* show. If the hypothesis is not supported, the scientist uses the observations from the experiment to make a new or revised hypothesis. Then, new experiments are planned.

Outcomes of the scientific method
Science is an iterative process in which hypotheses are developed and tested repeatedly. As new experimental results are obtained, old conclusions may need to be modified to fit new results. After significant experimentation, hypotheses with sufficient support may be elevated to the status of law or theory.

Law
A law is a statement of an order or relation of phenomena that, as far as is known, is invariable under the given conditions. Everything we observe in the universe operates according to known natural laws. If the truth of a statement is verified repeatedly in a reproducible way then it can reach the level of a natural law. Some well known and accepted natural laws of science are listed below:

        1.   The First Law of Thermodynamics

        3.   The Law of Cause and Effect

        4.   The Law of Biogenesis

Theory

In contrast to a law, a scientific theory is used to explain an observation or a set of observations. It is generally accepted to be true, though it cannot be proved absolutely. The important thing about a scientific theory is that there are no experimental observations to prove it NOT true, and each piece of evidence that exists supports the theory as written. Theories are often accepted at face value since they are often difficult to prove and can be rewritten in order to include the results of all experimental observations. An example of a theory is the big bang theory. While there is no experiment that can directly test whether or not the big bang actually occurred, there is no strong evidence indicating otherwise.

Theories provide a framework to explain the known information of the time, but are subject to constant evaluation and updating. There is always the possibility that new evidence will conflict with a current theory. Some examples of theories that have been rejected because they are now better explained by current knowledge are list below:

1. Theory of Spontaneous Generation
2. Inheritance of Acquired Characteristics
3. The Blending Hypothesis

Some examples of theories that were initially rejected because they fell outside of the accepted knowledge of the time but are well-accepted today due to increased knowledge and data include:

1. The sun-centered solar system
2. The germ theory of disease
3. Continental drift

Skill 2.2 **Analyzing ethical issues related to the process of doing science (e.g., accurately reporting experimental results)**

Ethical behavior is critical at all points of scientific experimentation, interpretation, and communication. Both other scientists and the public in general need to be confident in the validity of scientific results and the fact that science is being performed in the service of public good.

The first aspect of this is that **science is actually performed responsibly**. It is always important that careful documents be kept of all experimental conditions and outcomes. In both commercial and academic research environments, scientists keep careful records in laboratory notebooks. These serve as primary documents bearing witness to new discoveries and developments. Additionally, it is of utmost importance that, when used, animal subjects are humanely treated and human subjects are fully aware of all relevant facts (informed consent). Most research and funding organizations have policies in place and review boards to monitor the use of animal and human subjects.

Secondly, the **scientific findings must be accurately presented**. This means that all relevant data is presented to clearly reflect experimental findings. Occasionally, a scientist will report only that data which agrees with the hypothesis he is putting forth. Alternatively, the data may be improperly statistically manipulated. These and other misrepresentations of experimental findings seriously impede the progress of science because they may provide misinformation about how the natural world operates.

Another aspect of the requirement to accurately present research is the need for the **credit for work to be properly assigned**. Whether in the development of technology or in pure research, there are both financial and scientific issues at stake. In academic and pure research settings findings are typically published in peer reviewed journals and student theses. Occasionally, cases of plagiarism or even the theft of experimental work occur. Plagiarism means that the work has been presented as original when, in fact, it incorporates the work of others. Both these occurrences are highly undesirable and may results in disciplinary action and/or loss of reputation within the scientific community. It should be noted that, when preparing documents, it is acceptable to include information and even direction quotations from other sources as long as they are always *properly cited*.

One final area in which ethics is extremely important is in **deciding in what direction scientific research should proceed**. Both scientists and the general public must determine if there are any possible investigations that *should* not actually be performed. Currently, this is especially important in medical research where appropriateness of research into human cloning, embryonic stems cells, and other topics are heavily debated.

Skill 2.3 Evaluating the appropriateness of a specified experimental design to test a given physics hypothesis

Disturbance by observation

Experiments designed to test physics hypotheses are plagued by a number of difficulties that can be partially, but never totally, alleviated. For example, as has been pointed out by many, an experiment does not involve a scientist making an observation of a phenomenon; rather, it involves a scientist making an observation of a phenomenon being observed by a scientist. That is to say, the act of observation has an effect on the system being observed and this must be recognized in analyzing and reporting results. This effect can be profound in microscopic experiments, especially at the level of subatomic particles where using light to observe a particle actually disturbs the particle substantially. A macroscopic example is the measurement of voltage or current in an electrical circuit. Although most voltmeters and ammeters do not present a large load to the circuit, they do load the circuit to some degree affecting the measurements. Thus, it is critical to note the inherent limitations present in an experiment when evaluating its ability to properly test some hypothesis.

Isolation of parameters

Along similar lines to the above-mentioned difficulty, it is not possible to completely isolate a particular parameter for experimental testing. Thus, for example, although theory can treat a mass as the sole entity in the universe, in reality there are virtually innumerable other masses distributed throughout the universe that have an effect (by gravitation, for example) on the mass of interest. An appropriate experiment is thus designed to minimize outside and unwanted influences to the extent possible. One should evaluate how well the experiment tests only those parameters that are of concern to the hypothesis and how well the experiment reduces, or holds invariable, other parameters that could otherwise affect the results.

Measurement precision

Another concern regarding experimental design is the ability of the equipment to measure the desired parameters with sufficient accuracy and precision. If the parameter that is being measured is beyond the range of the equipment by way of being either too large, too small, or having variations that are too fine, the experiment can yield no useful information. The measurement equipment may be more or less technical; it may simply involve the use of the human eye or it may involve sophisticated equipment such as radiation or thermal energy detectors, pressure sensors or other devices.

Financial burden

Financial concerns are no less important than physical or scientific concerns. An experiment is only as useful as the possibility of its implementation and the cost of an experiment is a critical consideration in its appropriate design. High-energy physics, for example, is hampered by this difficulty as the costs of constructing particle accelerators and other associated equipment, not to mention the costs of maintenance and personnel, can be staggeringly high. Thus, an ideal experimental design is one that can be implemented at an affordable level of expense that allows for repeated use and proper maintenance.

Skill 2.4   Recognizing the role of communication among scientists and between scientists and the public in promoting scientific progress

There are many important reasons for scientists to report their detailed findings to the scientific community. The first and most important reason is to **ensure the correctness of scientific results and to advance our larger understanding of the natural world**. When new scientific findings are reported at conferences or peer reviewed for publication in technical journals, they are rigorously evaluated. This is to ensure that the experiments were performed with proper controls and that the results are repeatable. In controversial situations, experiments may even be fully repeated by other scientists to ensure that the same results are obtained. Once it is established that new findings are sound, the scientists can work together to determine how these results agree or disagree with previous experiments. They may also decide together what additional investigations would be useful to the field. **When open-minded scientific discourse is supported, opinions and information can be exchanged and theories can be refined.** Scientists will weigh the experimental evidence to determine what hypothesis has the most support. Given time, new theories may emerge and scientific knowledge can both broaden and deepen.

Science has increasingly become a collaborative and interdisciplinary effort. Engineers, clinicians, and scientists may not be familiar with advances in far reaching fields unless new developments are announced to a broad community of scientists. When this happens, it opens the doors for an exchange of information that can be enormously helpful in both investigating and solving problems in science and technology. Consider, for instance, an exchange of information between physicists and biologists. The physicists may suggest the use of high powered microscopes and lasers for study of sub-cellular structures. Likewise, the biologist's knowledge of transport proteins could inform the physicist's design of new nano-machines.

It is also important that the public in general be kept informed of scientific progress. This is of key importance in any democratic society because all people together must decide the direction of the government. **New scientific discoveries can have huge implications for medical, social, and environmental issues which should be of interest to all voters**. Additionally, when people are made familiar with the advances made in science, they may be more willing to see the need for public funding which supports continued scientific progress.

COMPETENCY 3.0 UNDERSTAND THE PROCESSES OF GATHERING, ORGANIZING, REPORTING, AND INTERPRETING SCIENTIFIC DATA AND APPLY THIS UNDERSTANDING IN THE CONTEXT OF PHYSICS INVESTIGATIONS

Skill 3.1 Evaluating the appropriateness of a given method or procedure for collecting data for a specified purpose

Resistance to human error

A number of factors influence the appropriateness of data collection methods with regard to the purpose for which they are being employed. The likelihood of human error must be screened carefully, as mistakes in setting up equipment, transcription errors in recording data or disturbances of the setup during data collection can all lead to skewed results. In some situations, it may be sufficient for the scientist to simply take measurements and record them by hand, but, at other times, a computer or data acquisition system may be required. In many electromagnetics applications, for example, the data rates or frequencies of operation are often so high as to make automated data collection a necessity.

Sufficient sampling

The data collection method must be able sample data at a sufficiently high rate to adequately characterize the process or phenomenon being studied. In the case of transient wave phenomena, such as transient electromagnetic or acoustic signals, the data collection rate must be at least twice as high as the highest frequency component of the signal. This so-called Nyquist rate is the minimum threshold for preventing aliasing in the collected data. The minimum requirements for data sampling may be temporal, as with many acoustic and electromagnetic signals, or they may be spatial, as with remote sensing, for example.

Sufficient data storage

Along similar lines, sufficient storage space for data must be available. Often, this involves computer storage in the form of a hard drive, or, perhaps, an optical medium. Data acquisition systems operating at high rates can produce tremendous amounts of data in short periods of time. Financial and even space considerations apply when a large number of CDs, DVDs or other media are required for storage.

Data precision

The precision of the data collection approach must also be evaluated. Both the sensing or measurement device, as well as the digitization equipment (if computer-based data acquisition is being used), must be of sufficient precision to measure the parameters of the system of interest to accurately record variations in those parameters. If, for example, a transducer is too coarse in its measurement abilities or an analog-to-digital converter does not have a sufficient number of bits to accurately digitize the signal, then information is lost. Similarly, the data that is collected should not be assigned more precision than the equipment allows. If the digitization equipment is more precise than the measurement or transduction equipment, for example, then the data collected may have more numerical precision than the transducer could have possibly afforded.

Skill 3.2 Selecting an appropriate and effective graphic representation (e.g., graph, table, diagram) for organizing, reporting, and analyzing given experimental data

In scientific investigations, it is often necessary to gather and analyze large data sets. We may need to manage data taken over long periods of time and under various different conditions. Appropriately organizing this data is necessary to identify trends and present the information to others. The uses and advantages of various graphic representations are discussed below.

Tables
Tables are excellent for organizing data as it is being recorded and for storing data that needs to be analyzed. In fact, almost all experimental data is initially organized into a table, such as in a lab notebook. Often, it is then entered into tables within spreadsheets for further processing. Tables can be used for presenting data to others if the data set is fairly small or has been summarized (for example, presenting average values). However, for larger data sets, tabular presentation may be overwhelming. Finally, tables are not particularly useful for recognizing trends in data or for making them apparent to others.

Charts and Graphs

Charts and graphs are the best way to demonstrate trends or differences between groups. They are also useful for summarizing data and presenting it. In most types of graphs, it is also simple to indicate uncertainty of experimental data using error bars. Many types of charts and graphs are available to meet different needs. Three of the most common are scatterplots, bar charts, and pie charts. An example of each is shown.

Scatterplots are typically shown on a Cartesian plane and are useful for demonstrating the relationship between two variables. A line chart (shown) is a special type of two-dimensional scatter plot in which the data points are connected with a line to make a trend more apparent.

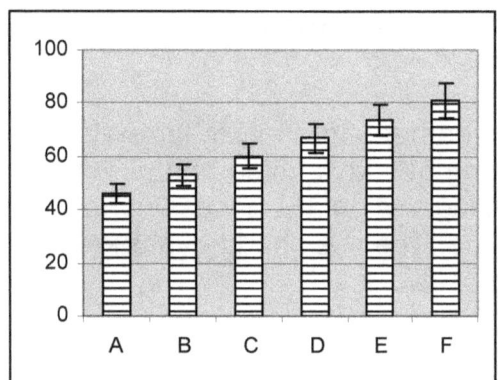

Bar charts can sometimes fill the same role as scatterplots but are better suited to show values across different categories or different experimental conditions (especially where those conditions are described qualitatively rather than quantified). Note the use of error bars in this example.

Finally, a pie chart is best used to present relative magnitudes or frequencies of several different conditions or events. They are most commonly used to show how various categories contribute to a whole.

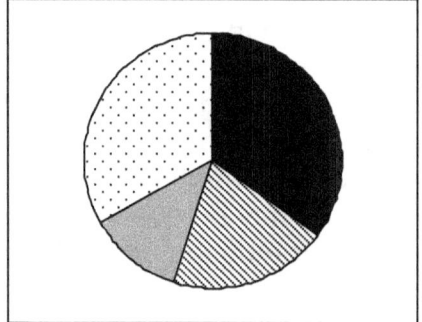

Diagrams

Diagrams are not typically used to present the specifics of data. However, they are very good for demonstrating phenomena qualitatively. Diagrams make it easy to visualize the connections and relationships between various elements. They may also be used to demonstrate temporal relationships. For example, diagrams can be used to illustrate the operation of an internal combustion engine or the complex biochemical pathways of an enzyme's action. The diagram shown is a simplified version of the carbon cycle.

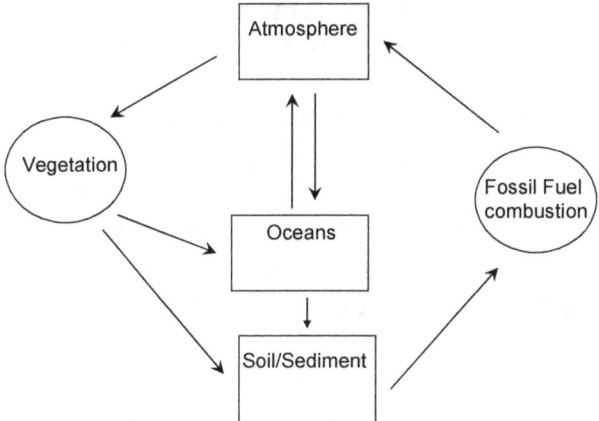

Skill 3.3 Applying procedures and criteria for formally reporting experimental procedures and data to the scientific community

Once scientists have rigorously performed a set of experiments, they may believe that they have information of significant value that should be shared within the scientific community. **New findings are often presented at scientific conferences and ultimately published in technical journals** (some well known examples include *Nature*, *Science*, and the *Journal of the American Medical Association*). Scientists prepare manuscripts detailing the conditions of their experiments and the results they obtained. They will typically also include their interpretation of those results and their impact on current theories in the field. These manuscripts are not wholly unlike lab reports, though they are considerably more polished, of course. Manuscripts are then submitted to appropriate technical journals.

All reputable scientific journals use peer review to assess the quality of research submitted for publication. Peer review is the process by which scientific results produced by one person or group are subjected to the analysis of other experts in the field. It is important that they be objective in their evaluations. Peer review is typically done anonymously so that the identity of the reviewer remains unknown by the scientists submitting work for review. The goal of peer review is to "weed out" any science not performed by appropriate standards. Reviewers will determine whether proper controls were in place, enough replicates were performed, and that the experiments clearly address the presented hypothesis. The reviewer will scrutinize the interpretations and how they fit into what is already known in the field. Often reviewers will suggest that additional experiments be done to further corroborate presented conclusions. If the reviewers are satisfied with the quality of the work, it will be published and made available to the entire scientific community.

Sometimes, there is significant opposition to new ideas, especially if they conflict with long-held ideas in a scientific field. However, with enough correct experimental support, scientific theories will in time be adjusted to reflect the newer findings. Thus science is an ongoing cycle: ideas are constantly being refined and modified to reflect new evidence and, ultimately, provide us with a more correct model of the world around us. There are many scientific theories that have experienced multiple revisions and expansions. Some well-known examples include the atomic theory, which was notable refined by Dalton, Thomson, Rutherford, and Bohr and the theory of natural selection, which was originally formulated by Darwin but has been enriched by discoveries in molecular biology.

Skill 3.4 Analyzing relationships between factors (e.g., linear, direct, inverse, direct squared, inverse squared) as indicated by experimental data

In many experimental investigations we attempt to understand the relationship between two variables. Typically, one variable (independent, usually represented x) is altered while the other is measured (dependent, usually represented y). Certain relationships between these variables are commonly observed. Often, the relationships are most clearly observable when graphed.

Linear
In linear relationships, the independent variable is *directly* proportional to the dependent variable. This means that, as the absolute value of x is increased, the absolute value of y will also increase. This is stated mathematically as:

$$y \propto x$$

This relationship can also be represented using the familiar equation for a straight line (where m is slope and b is the intercept):

$$y = mx + b$$

Graphs of this type of relationship are shown below. On the left, the slope of the line (m) is positive and on the right it is negative:

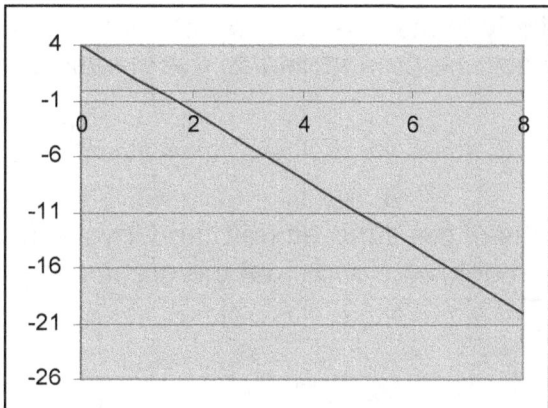

Inverse

Unlike direct proportionality, inverse proportionality signifies that the dependent variable is a function of the inverse of the independent variable:

$$y \propto \frac{1}{x}$$

A graph of this relationship is hyperbolic, as shown below:

Squared and Inverse Squared

As the name suggests, in squared relationships the dependent variable is proportional to the square of the independent variable:

$$y \propto x^2$$

Similarly to the inverse proportionality described above, the dependent variable may also be proportional to the inverse square of the independent variable:

$$y \propto \frac{1}{x^2}$$

Graphs of the squared (left) and inversely squared (right) proportionality are shown below. Notice that the graphs are non-linear.

 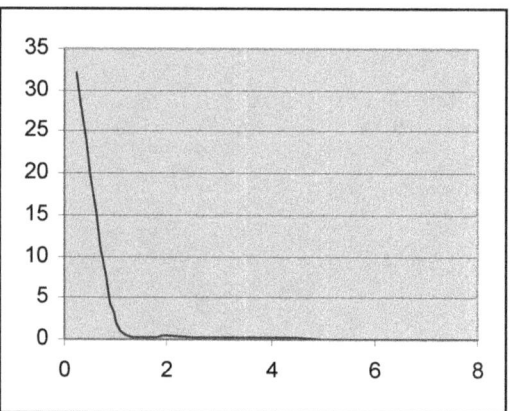

Logarithmic and Exponential

The dependent variable may also be proportional to a logarithmic or exponential function of the independent variable.

$$y \propto \log_a x \qquad y \propto a^x$$

Graphs of these relationships are shown below (logarithmic relationship on the left and exponential on the right). Make special note of the scale of the dependent variable.

Skill 3.5 **Coordinating explanations at different levels of scale, points of focus, and degrees of complexity and specificity**

Physics has shown, especially through its revolutionary period in the early 20th century, that theories and explanations of phenomena must be treated carefully relative to how broadly and how fundamentally they describe those phenomena. For example, Newtonian mechanics provided, and still provides, a useful and accurate theory for studying the motion of macroscopic objects. The rules governing the behavior of the subatomic particles that constitute these macroscopic objects, however, are fundamentally different. This shows that **objects at different scales cannot always be treated using the same set of principles or theories**. Nevertheless, it can sometimes be shown that one theory is actually a subset of another theory. The correspondence principle states that quantum mechanics and classical (or Newtonian) mechanics converge in the limit of large quantum numbers.

It is often very useful to limit the complexity of an analysis based on the complexity of the data needed. Thus, **if a simple approach is available and provides sufficiently accurate results, there may be no need to take a more complicated approach**. Due to the complexity of many systems, it is often impossible to perform an analysis that treats every aspect of the system exactly. Nevertheless, it is possible to analyze such systems by considering an ensemble of particles or other features that are more manageable.

Statistical analysis is one way to approach this problem. Semiconductors involve a large number of subatomic particles that must be treated quantum mechanically and solving the Schrödinger equation for even a few particles is difficult. Thus, typically, the behavior of small sets (or regularly ordered sets) of particles are calculated and the principles generalized to cases where large numbers of particles are involved.

Based on the examples of semiconductors and quantum mechanics, it can be seen that **explanations of phenomena in one area are not always applicable to similar phenomena in other areas**. Quantum mechanics suggests that subatomic particles cannot be treated like billiard balls. On the other hand, within a certain range, Newtonian mechanics is indeed applicable regardless of scale: planets may be treated in much the same manner as billiard balls, for example.

Care must then be taken with regard to the use of theories. Empirically-based theories, for example, are built upon a range of observations from a certain set of contingent circumstances. Simply because interpolation among these data may provide accurate predictive power does not necessarily imply that extrapolation beyond the initial range of circumstances can provide equally accurate predictive power. Again, this is illustrated by the relationship of quantum and Newtonian mechanics. **Every theory attempts to provide an explanation of certain phenomena at a certain scale and to a certain level of complexity, and it is only with extreme caution that a theory can be extended beyond these self-imposed boundaries.**

Skill 3.6   Applying statistical techniques to analyze data

Measurements of data are subject to numerous types of errors (or "noise") that can lead to variations in results that might otherwise not have been expected. These **variations in measurement data can result from inherent noise in equipment, human error or other factors**. Since no one measurement or datum can generally be treated as any more correct than another measurement or datum, as the variations are often random, the aggregate data must be treated statistically.

The effect of the use of statistics in analyzing data is to remove or quantify factors that adversely affect the precision of a set of measurements or other data. Alternatively, statistics can be used to provide a more compact description of a set of data corresponding to random variables.

An example experiment is the rolling of a ball on a flat surface: the ball is launched with a certain force and the distance traveled is measured to gain some insight into friction. The ball will likely travel a different distance for each trial, although the variation may be small. These differences can result from an inability of the launching device (perhaps the human hand) to provide the same impulse for each instance, from small, random variations in the texture of the surface, or from measurement error during the determination of the distance traveled. **Analysis of the aggregate data by calculating a mean or median value and the variance or standard deviation of the measurements, for example, can provide a more informative assessment of the results than can any particular measurement datum**.

For a set of n data values $x_1, x_2, ... x_n$, some of the common statistical parameters calculated are:

Mean given by $\bar{x} = \dfrac{\sum x_i}{n}$

Median or middle value in an ordered list of data
Mode or most frequently occurring value
Standard deviation given by $\sqrt{\dfrac{\sum (x_i - \bar{x})^2}{n}}$

For sufficiently large measurement (or other data) populations, **statistical distributions** can be created to make probabilistic predictions about future measurement results or responses of a system. This is crucial for such situations as, for example, probability of failure for some system. The accuracy or precision of a measurement device can also be treated statistically.

Proper statistical analysis of data is crucial, especially when that data is used to calculate other parameters. Error propagation (or uncertainty propagation) must be treated carefully, as the variation in a measured parameter, such as the position of a moving car, cannot be directly translated into the variation of a calculated parameter, such as the acceleration of that car. The error, or uncertainty, of the acceleration must be appropriately calculated based on the error of the position measurements; the two errors are generally not the same value.

Statistical analysis can be used for complex systems when exact treatment of each individual aspect of the system is impossible. **A statistical analysis allows for the determination of the behavior of a system on average, thus providing some insight when an exact analysis is too costly in terms of time or other resources**.

TEACHER CERTIFICATION STUDY GUIDE

COMPETENCY 4.0 UNDERSTAND PRINCIPLES AND PROCEDURES OF MEASUREMENT AND THE SAFE AND PROPER USE OF EQUIPMENT AND MATERIALS USED IN PHYSICS INVESTIGATIONS

Skill 4.1   Evaluating the appropriateness of SI units of measurement, measuring devices, or methods of measurement for given situations

In accordance with the popular saying, "the right tool for the right job," experiments require the proper equipment and approaches for measurement to be successful. A number of factors ranging from what is being measured to the context in which measurements are being taken affect how an experiment is to be conducted. Particular fields or industries, even, may have specific conventions for measurements.

SI Units

The International System of Units, or SI units, is the most common system of units used in the physical sciences. The abbreviation SI originates from the French name, *Le Système International d'Unités*. SI units are commonly accepted among most scientists for collecting and reporting measurements. In a few cases there are some exceptions with regard to SI unit usage. For instance, in the heating and air conditioning industries, the British thermal unit (BTU) is used in place of the Joule, the SI unit for energy. A system of units is always based on certain conventions, such as the precise definition of a kilogram, a meter or a second, and many other units, such as the Newton (unit of force) or the Joule (unit of energy) are derived from these definitions. No particular system of units is inherently better than another although some may be more convenient to use. SI is often a preferred system due to its ubiquity among scientists, its largely base-10 systematization and its popular use by most countries around the world. Some scientists still use the centimeter-gram-second (CGS) system of units due to its usefulness in simplifying some equations and in reducing some physical constants to unity.

Measurement devices and methods

The approach to taking measurements for a particular experiment is determined by the particular phenomenon being measured, the context of measurement and the financial limitations imposed upon the experiment. Obviously, thermometers cannot be used for measuring distances, and the right equipment must be used for the appropriate measurement. Nevertheless, more subtle considerations abound, including the required accuracy and precision of measurements. If relatively coarse measurements are sufficient, there is often no need for sophisticated, expensive equipment.

For example, if a rough measurement of weight is required, a mechanical balance scale may be all that is required. Measurements of higher precision using an advanced digital scale may be gratuitous, especially in light of other inexact measurements that may be taken during the experiment.

The particular method used for measurement is determined in large part by the theory or hypothesis being tested. Measurements in the realm of particle physics, where hypotheses are largely based on a synthesis of quantum mechanics and special relativity, do not allow for the use of approaches to measurement that assume classical mechanics and electrodynamics. Instead, the particular method used must be based on more firmly established principles and concepts from the theory. Additionally, certain types of measurements must be approached indirectly. That is to say, empirical information about some parameters of a system may actually require measurement of different parameters from which measurements the desired parameter can be calculated. Such indirect measurements may be required for phenomena that are newly discovered, that are microscopic in scale or that require extremely high-precision results.

Skill 4.2 Analyzing likely sources of error in given measurements in physics experiments

Skill 3.6 discusses the sources of error in measurement and how they are handled statistically. Here we will analyze the propagation of error and accurate estimation of error in a given situation.

Often in scientific operations we want to determine a quantity that requires many steps to measure. Of course, each time we take a measurement there will be a certain associated error that is a function of the measuring device. Each of these errors contributes to an even greater one in the final value. This phenomenon is known as **propagation of error or propagation of uncertainty**.

A measured value is typically expressed in the form $x \pm \Delta x$, where Δx is the uncertainty or margin of error. What this means is that the value of the measured quantity lies somewhere between $x - \Delta x$ and $x + \Delta x$, but our measurement techniques do not allow us any more precision. If several measurements are required to ultimately decide a value, we must use formulas to determine the total uncertainty that results from all the measurement errors. A few of these formulas for simple functions are listed below:

Formula   Uncertainty

$$X = A \pm B \qquad (\Delta X)^2 = (\Delta A)^2 + (\Delta B)^2$$

$$X = cA \qquad \Delta X = c\Delta A$$

$$X = c(A \cdot B) \qquad \left(\frac{\Delta X}{X}\right)^2 = \left(\frac{\Delta A}{A}\right)^2 + \left(\frac{\Delta B}{B}\right)^2$$

$$X = c\left(\frac{A}{B}\right) \qquad \left(\frac{\Delta X}{X}\right)^2 = \left(\frac{\Delta A}{A}\right)^2 + \left(\frac{\Delta B}{B}\right)^2$$

For example, if we wanted to determine the density of a small piece of metal we would have to measure its weight on a scale and then determine its volume by measuring the amount of water it displaces in a graduated cylinder. There will be error associated with measurements made by both the scale and the graduated cylinder. Let's suppose we took the following measurements:

Mass: 57 ± 0.5 grams
Volume: 23 ± 3 mm³

Since density is simply mass divided by the volume, we can determine its value to be:

$$\rho = \frac{m}{V} = \frac{57g}{23mm^3} = 2.5\frac{g}{mm^3}$$

Now we must calculate the uncertainty on this measurement, using the formula above:

$$\left(\frac{\Delta x}{x}\right)^2 = \left(\frac{\Delta A}{A}\right)^2 + \left(\frac{\Delta B}{B}\right)^2$$

$$\Delta x = \left(\sqrt{\left(\frac{\Delta A}{A}\right)^2 + \left(\frac{\Delta B}{B}\right)^2}\right) x = \left(\sqrt{\left(\frac{0.5g}{57g}\right)^2 + \left(\frac{3mm^3}{23mm^3}\right)^2}\right) \times 2.5\frac{g}{mm^3} = 0.3\frac{g}{mm^3}$$

Thus, the final value for the density of this object is 2.5 ± 0.3 g/mm³.

PHYSICS

Skill 4.3 Distinguishing between accuracy and precision in scientific measurements

Accuracy is a measure of how close to "correct" a measuring device or technique is. Precision is a measure of how similar repeated measurements from a given device or technique are. While the best devices or techniques will yield measurements that are both accurate and precise, it is possible to be accurate without being precise or to be precise without being accurate. The classic analogy to demonstrate accuracy and precision is that of a bulls-eye. Accuracy alone is shown in the left example: the shots are all close to the center of the bulls-eye (the correct value). Precision alone is shown in the middle example: the shots are tightly clustered together. Both accuracy and precision are shown in the example on the right: the shots are tightly clustered near the center of the bulls-eye.

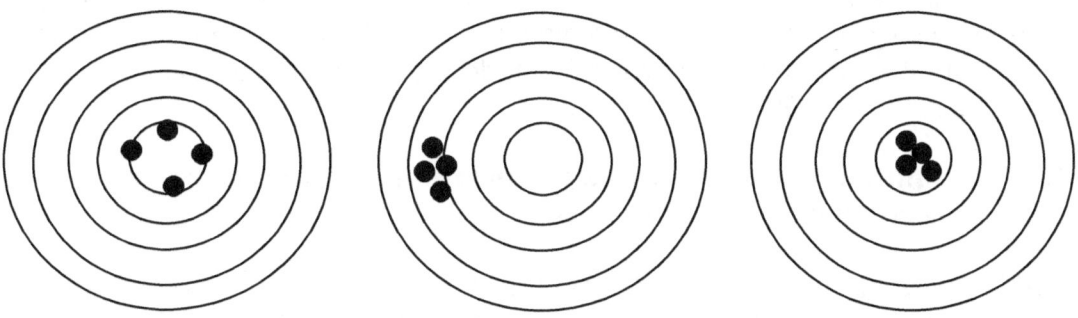

The accuracy of a technique or device can be determined in a straightforward manner. It is done by measuring a known quantity (a standard) and determining how close the value provided by the measuring device is.

To determine precision, we must make multiple measurements of the same sample. The precision of an instrument is typically given in terms of its standard error or standard deviation. Precision is typically divided into reproducibility and repeatability. These concepts are subtly different and are defined as follows:

Repeatability: Variation observed in measurements made over a short period of time while trying to keep all conditions the same (including using the same instrument, the same environmental conditions, and the same operator)

Reproducibility: Variation observed in measurements taken over a long time period in a variety of different settings (different places and environments, using different instruments and operators)

Both repeatability and reproducibility can be estimated by taking multiple measurements under the conditions specified above. Using the obtained values, standard deviation can be calculated using the formula:

$$\sigma = \sqrt{\frac{1}{N}\sum_{i=1}^{N}(x_i - \overline{x})^2}$$

where σ = standard deviation
N = the number of measurements
x_i = the individual measured values
$\overline{x}$ = the average value of the measured quantity

To obtain a reliable estimate of standard deviation, N, the number of samples, should be fairly large. We can use statistical methods to determine a confidence interval on our measurements. A typical confidence level for scientific investigations is 90% or 95%.

Skill 4.4   Using significant figures in performing calculations and expressing measurements

Significant figures are important to denote the degree of certainty we can have in a given measurement or calculation. Rules exist to allow us to first calculate the number of significant figures in a given quantity and then to determine how many significant figures should remain following an operation. First, how to determine the number of significant figures in a quantity:

1. Non-zero digits are always significant.
2. Any zeros between two significant digits are significant.
3. A final zero or trailing zeros in the decimal portion ONLY are significant.

For instance, the number of significant figures in the following measurements are as follows:

1) 3.0800   5 significant figures

2) 0.00418   3 significant figures

3) 7.09×10^{-5} 3 significant figures

4) 91,600   3 significant figures

Now the rules for operations:

1. For multiplication or division, keep the same number of significant figures as the factor with the *fewest* significant figures.

2. For addition or subtraction, keep the same number of decimal places as the term with the *fewest*.

Here are examples of each of these rules:

1) 1.2 x 4.56 = 5.472 but since the first factor has only 2 significant figures, the answer must be also have 2 significant figures and is 5.5

1.234 + 5.67 = 6.904 but since the second term has only 2 decimal places, so must the answer which is 6.90

Skill 4.5 Analyzing the principles upon which given laboratory instruments are based (e.g., oscilloscopes, Geiger counters)

Oscilloscope: An oscilloscope is a piece of electrical test equipment that allows signal voltages to be viewed as two-dimensional graphs of electrical potential differences plotted as a function of time.

The oscilloscope functions by measuring the deflection of a beam of electrons traveling through a vacuum in a cathode ray tube. The deflection of the beam can be caused by a magnetic field outside the tube or by electrostatic energy created by plates inside the tube. The unknown voltage or potential energy difference can be determined by comparing the electron deflection it causes to the electron deflection caused by a known voltage.

Oscilloscopes can also determine if an electrical circuit is oscillating and at what frequency. They are particularly useful for troubleshooting malfunctioning equipment. You can see the "moving parts" of the circuit and tell if the signal is being distorted. With the aid of an oscilloscope you can also calculate the "noise" within a signal and see if the "noise" changes over time.

Inputs of the electrical signal are usually entered into the oscilloscope via a coaxial cable or probes. A variety of transducers can be used with an oscilloscope that enable it to measure other stimuli including sound, pressure, heat, and light.

Voltmeter, Ammeter and other electrical meters: See Skill 13.5.

Signal Generator: A signal generator, also known as a test signal generator, function generator, tone generator, arbitrary waveform generator, or frequency generator, is a device that generates repeating electronic signals in either the analog or digital domains. They are generally used in designing, testing, troubleshooting, and repairing electronic devices.

A function generator produces simple repetitive waveforms by utilizing a circuit called an electronic oscillator or a digital signal processor to synthesize a waveform. Common waveforms are sine, sawtooth, step or pulse, square, and triangular. Arbitrary waveform generators are also available which allow a user to create waveforms of any type within the frequency, accuracy and output limits of the generator. Function generators are typically used in simple electronics repair and design where they are used to stimulate a circuit under test. A device such as an oscilloscope is then used to measure the circuit's output.

Spectrometer: A spectrometer is an optical instrument used to measure properties of light over a portion of the electromagnetic spectrum. Light intensity is the variable that is most commonly measured but wavelength and polarization state can also be determined. A spectrometer is used in spectroscopy for producing spectral lines and measuring their wavelengths and intensities. Spectrometers are capable of operating over a wide range of wavelengths, from short wave gamma and X-rays into the far infrared. In optics, a spectrograph separates incoming light according to its wavelength and records the resulting spectrum in some detector. In astronomy, spectrographs are widely used with telescopes.

Geiger counter: A Geiger counter detects ionizing radiation, i.e. radiation capable of ionizing atoms and molecules and causing damage to DNA and living tissue. The core of a Geiger counter is a Geiger-Müller tube, a tube filled with inert gas (typically helium, neon or argon) and containing a thin metal wire carrying a charge. When a nuclear particle penetrates the tube, it temporarily makes the gas conductive and electrons are ultimately attracted to the central conducting wire. As the particle travels towards the wire, it continues to collide with atoms and release more electrons, thereby amplifying the signal. Ultimately, a pulse is produced by each particle that enters the detector. The counters are designed to display this signal with a needle or lamp or to produce audible clicks.

The earliest device of this type was produced by Ernest Rutherford in 1908 and the design was later refined by Hans Geiger and Walther Müller 20 years later. They are presently called "counters" because every particle produces a pulse or click, allowing the total number of particles to be tabulated. Geiger counters are commonly used because they are sturdy and relatively cheap to produce. However, they are only capable of determining the intensity of radiation (number of particles present over a given time). The energy level of the particles is also an important quantity but basic Geiger counters are not able to measure this.

Skill 4.6 Analyzing hazards associated with given laboratory materials (e.g., projectiles, lasers, radioactive materials, heat sources, chemicals)

Motion and forces: All stationary devices must be secured by C-clamps. Protective goggles must be used. Care must be taken at all times while knives, glass rods and heavy weights are used. Viewing a solar eclipse must always be indirect. When using model rockets, NASA's safety code must be implemented.

Heat: The master gas valve must be off at all times except while in use. Goggles and insulated gloves are to be used whenever needed. Never use closed containers for heating. Burners and gas connections must be checked periodically. Gas jets must be closed soon after the experiment is over. Fire retardant pads and quality glassware such as Pyrex must be used.

Pressure: While using a pressure cooker, never allow pressure to exceed 20 lb/square inch. The pressure cooker must be cooled before it is opened. Care must be taken when using mercury since it is poisonous. A drop of oil on mercury will prevent the mercury vapors from escaping.

Light: Broken mirrors or those with jagged edges must be discarded immediately. Sharp-edged mirrors must be taped. Spectroscopic light voltage connections must be checked periodically. Care must be taken while using ultraviolet light sources. Some students may have psychological or physiological reactions to the effects of strobe like (e.g. epilepsy).

Lasers: Direct exposure to lasers must not be permitted. The laser target must be made of non-reflecting material. The movement of students must be restricted during experiments with lasers. A number of precautions while using lasers must be taken – use of low power lasers, use of approved laser goggles, maintaining the room's brightness so that the pupils of the eyes remain small. Appropriate beam stops must be set up to terminate the laser beam when needed. Prisms should be set up before class to avoid unexpected reflection.

Sound: Fastening of the safety disc while using the high speed siren disc is very important. Teacher must be aware of the fact that sounds higher than 110 decibels will cause damage to hearing.

Radiation: Proper shielding must be used while doing experiments with x-rays. All tubes that are used in a physics laboratory such as vacuum tubes, heat effect tubes, magnetic or deflection tubes must be checked and used for demonstrations by the teacher. Cathode rays must be enclosed in a frame and only the teacher should move them from their storage space. Students must watch the demonstration from at least eight feet away.

TEACHER CERTIFICATION STUDY GUIDE

Radioactivity: The teacher must be knowledgeable and properly trained to handle the equipment and to demonstrate. Proper shielding of radioactive material and proper handling of material are absolutely critical. Disposal of any radioactive material must comply with the guidelines of NRC.

Chemicals: All laboratory solutions should be prepared as directed in the lab manual. Care should be taken to avoid contamination. All glassware should be rinsed thoroughly with distilled water before using and cleaned well after use. Unused solutions should be disposed of according to local disposal procedures. Chemicals should not be stored on bench tops or heat sources. They should be stored in groups based on their reactivity with one another and in protective storage cabinets. All containers within the lab must be labeled. Chemical waste should be disposed of in properly labeled containers. Waste should be separated based on their reactivity with other chemicals. **Material safety data sheets** are available for every chemical substance directly from the company of acquisition or the internet.
The following chemicals are potential carcinogens and not allowed in school facilities: Acrylonitriel, Arsenic compounds, Asbestos, Bensidine, Benzene, Cadmium compounds, Chloroform, Chromium compounds, Ethylene oxide, Ortho-toluidine, Nickel powder, and Mercury.

It is important that teachers and educators follow these guidelines to protect the students and to avoid most of the hazards. They have a responsibility to protect themselves as well. **There should be not any compromises in issues of safety.**

Skill 4.7 Applying safety rules regarding electricity and electrical equipment

Active learning about electromagnetism may require the use of live electrical sources and many physics laboratories can be greatly enhanced with the use of electrical devices. However, care must always be taken when dealing with electricity. The danger in working with electrical device is two-fold as follows.

1. Risk of electrical shock due to:
    a. Exposed electrical contacts
    b. Poor or damaged insulation in equipment or on cords
    c. Moisture in the area of electrical equipment

2. Risk of fire due to:
    a. Over-heating of equipment or overloaded circuits
    b. Faulty connections or short circuits
    c. Flammable substance in the area of electrical equipment

PHYSICS

Adherence to the following rules will minimize the risk of electrical accidents:
1. Experiments should never be performed by a single person working alone.
2. Check all equipment, switches, and leads one by one. If missing insulation, frayed cords, blackening (due to arcing), bent or missing prongs are in evidence, do not use the equipment.
3. Always connect and turn the power supply on last. When concluding work, turn the power supply off first.
4. Do not run wires (or extension cords) over moving or rotating equipment, or on the floor, or string them across walkways from bench-to-bench.
5. Do not wear conductive watch bands or chains, finger rings, wrist watches, etc., and do not use metallic pencils, metal or metal edge rulers when working with exposed circuits.
6. If breaking an inductive circuit open, turn your face away to avoid danger from possible arcing.
7. If using electrolytic capacitors, always wait an appropriate period of time (usually approximately five time constants) for capacitor discharge before working on the circuit.
8. All conducting surfaces serving as "ground" should be connected together.
9. All instructional laboratories should be equipped with Ground Fault Current Interrupt (GFCI) circuit breakers. If breakers repeatedly trip and overload is not present, check for leakage paths to ground.
10. Use only dry hands and tools, work on a dry surface, and stand on a dry surface.
11. Never put conductive metal objects into energized equipment.
12. Always carry equipment by the handle and/or base, never by the cord.
13. Unplug cords by pulling on the plug, never on the cord.
14. Use extension cords only temporarily. The cord should be appropriately rated for the job.
15. Use extension cords with 3 prong plugs to ensure that equipment is grounded.
16. Never remove the grounding post from a 3 prong plug.
17. Do not overload extension cords, multi-outlet strips, or wall outlets.

Skill 4.8 **Applying proper procedures for dealing with accidents and injuries in the physics laboratory**

Safety is a learned behavior and must be incorporated into instructional plans. Measures of prevention and procedures for dealing with emergencies in hazardous situations have to be in place and readily available for reference. Copies of these must be given to all people concerned, such as administrators and students.

The single most important aspect of safety is planning and anticipating various possibilities and preparing for the eventuality. Any Physics teacher/educator planning on doing an experiment must try it before the students do it. In the event of an emergency, quick action can prevent many disasters. The teacher/educator must be willing to seek help at once without any hesitation because sometimes it may not be clear that the situation is hazardous and potentially dangerous. Accidents and injuries should always be reported to the school administration and local health facilities. The severity of the accident or injury will determine the course of action to pursue.

There are a number of procedures to prevent and correct any hazardous situation. There are several safety aids available commercially such as posters, safety contracts, safety tests, safety citations, texts on safety in secondary classroom/laboratories, hand books on safety and a host of other equipment. Another important thing is to check the laboratory and classroom for safety and report it to the administrators before staring activities/experiments.

COMPETENCY 5.0 UNDERSTAND THE USE OF MATHEMATICS (E.G., DIMENSIONAL ANALYSIS, ALGEBRA, VECTOR ANALYSIS, CALCULUS) AND MATHEMATICAL MODELING IN PHYSICS

Skill 5.1 Using mathematics to derive and solve equations

Mathematics is a very broad field, encompassing various specific disciplines including calculus, trigonometry, algebra, geometry, complex analysis and other areas. **Many physical phenomena can be modeled mathematically using functions to relate a specific parameter or state of the system to one or more other parameters**. For example, the net force on a charged object is a function of the direction and magnitude of the forces acting upon it, such as electrical attraction or repulsion, tension from a string or spring, gravity and friction. When a phenomenon has been modeled mathematically as a set of expressions or equations, the general relationships of numbers can be applied to glean further information about the system for the purposes of greater understanding or prediction of future behavior.

The process of treating a system mathematically can involve use of empirical relationships (such as Ohm's law) or more *a priori* relationships (such as the Schrödinger equation). Given or empirically derived equations can be manipulated or combined, depending on the specific situation, to isolate a specific parameter as a function of other parameters (solution of the equation). This process can also lead to a different equation that presents a new or simplified relationship among specific parameters (derivation of an equation).

Both derivation and solution of equations can be pursued in either an **exact** or **approximate** manner. **In some cases, equations are intractable and certain assumptions must be made to facilitate finding a solution**. These assumptions are typically drawn from generalizations concerning the behavior of a particular system and will result in certain restrictions on the validity and applicability of a solution. An exact solution, presumably, has none of these limitations.

Another variation of approximate derivation and solution of an equation involves numerical techniques. Ideally, an analytical approach that employs no approximations is the best alternative; such an approach yields the broadest and most useful results. Nevertheless, the intractability of an equation can lead to the need for either an approximate or numerical solution. One particular example is from the field of electromagnetics, where determination of the scattering of radiation from a finite circular cylinder is either extremely difficult or impossible to perform analytically. Thus, either numerical or approximate approaches are required. The range of numerical techniques available depends largely on the type of problem involved. The approach for numerically solving simple algebraic equations, for example, is different from the approach for numerically solving complex integral equations.

Care must be taken with analytical (and numerical or approximate) approaches to physical situations **as it is possible to produce mathematically valid solutions that are physically unacceptable.** For instance, the equations that describe the wave patterns produced by a disturbance of the surface of water, upon solution, may yield an expression that includes both incoming and outgoing waves. Nevertheless, in some situations, there is no physical source of incoming waves; thus, the analytical solution must be tempered by the physical constraints of the problem. In this case, one of the solutions, although mathematically valid, must be rejected as unphysical.

Skill 5.2 Applying algebra and geometry to model physical situations

Geometry and algebra are key tools in modeling physical situations in a number of specific areas. Understanding physical systems involves examination of certain parameters of the system as some other parameter or set of parameters, such as time or a spatial dimension, is varied. Through measurements, these parameters can be quantified and plotted graphically. In many cases, the graph of the results of such an analysis will reveal a curve that relates the state of the system, in terms of some parameter, to the value of another parameter. Such a curve could be the position of a moving car with respect to time. **The empirical curve can then be fitted to an algebraic expression that expresses a mathematical relationship between the parameters by way of a function.** Thus, for the example of the traveling car, the position x of the car can be expressed as a function of time, $x(t)$. For a car moving at constant speed v, the function is of the form $x(t) = v t + x_0$, where x_0 is an initial position at time $t = 0$. A similar approach can be used to model more complicated situations.
Alternatively, other algebraic equations can be combined with each other, or with measured results, to derive new expressions. For empirically-based equations, curve fitting of various forms (such as least squares regression) can be employed, depending on the complexity of the data.

Algebra for physical situations may involve several different ranges or types of numbers. In some cases, the equations may involve **real numbers** (such as in time-domain analysis in classical electrodynamics), or they may involve **complex numbers** (such as in frequency-domain analysis in classical electrodynamics). Furthermore, the algebra may be **discrete** or **continuous**. Although there is some overlap, classical mechanics is based on the use of a continuous numerical scheme for energy states but quantum mechanics involves a discrete numerical scheme for energy states.

Geometry is another key component in modeling physical situations and can apply to either purely spatial parameters or to a combination of spatial, temporal or other parameters. For example, the forces acting on a charged object must be dealt with geometrically when a net force is to be determined. The force due to gravity, the force due to electrical attraction or repulsion and the force due to tension from a spring or string may all be involved. These forces must be treated vectorially and an understanding of the geometrical relationships is fundamental to an understanding of a behavior of the system. If the system is not at steady state, time becomes a factor that can enter algebraically as it affects the geometrical properties of the system.

The particular approach to geometry, in terms of the system of coordinates, depends largely on the characteristics of the problem. Equations that are extremely complex in rectangular coordinates, for example, may be quite simple in spherical coordinates. Commensurately, problems that involve primarily planes, for example, are best treated in rectangular coordinates.

Thus, although algebra and geometry may seem to be simple aspects of mathematical modeling of physical situations, they are in fact highly nuanced and can add considerable complexity and detail to models.

TEACHER CERTIFICATION STUDY GUIDE

Skill 5.3   Applying dimensional analysis to solve problems

Dimensional analysis is simply a technique in which the units of the variables in an equation are analyzed. It is often used by scientists and engineers to determine if a derived equation or computation is plausible. While it does not guarantee that a stated relationship is correct, it does tell us that the relationship is at least reasonable. When there is a mix of various physical quantities being equated in a relationship, the units of both sides of the equation must be the same. We will examine a simple example to demonstrate how dimensional analysis can be used:

The Ideal Gas Law is used to predict change in temperature, volume, or pressure of a gas. It is:

$$PV = nRT$$

Where P=pressure [Pa]
V=volume [m^3]
n=numbers of moles of gas [mol]
T=temperature [K]
R=the gas constant 8.314472 [m^3·Pa·K^{-1}·mol^{-1}]

Show that this formula is physically plausible using dimensional analysis.

Begin by substituting the units of each quantity into the equation:

$$[Pa] \times [m^3] = [mol] \left[\frac{[m^3] \times [Pa]}{[K] \times [mol]} \right] [K]$$

On the right-hand side of the equation, we can cancel [mol] and [K]:

$$[Pa] \times [m^3] = [m^3] \times [Pa]$$

Because this is a simple example, we can already see that the two sides of the equations have the same units. In more complicated equations, additional manipulation may be required to elucidate this fact. Now, we can confidently believe in the plausibility of this relationship.

TEACHER CERTIFICATION STUDY GUIDE

Skill 5.4   Applying trigonometric functions and graphing to solve problems (including vector problems)

Vector space is a collection of objects that have magnitude and direction. They may have mathematical operations, such as addition, subtraction, and scaling, applied to them. Vectors are usually displayed in boldface or with an arrow above the letter. They are usually shown in graphs or other diagrams as arrows. The length of the arrow represents the magnitude of the vector while the direction in which it points shows the direction.

To add two vectors graphically, the base of the second vector is drawn from the point of the first vector as shown below with vectors **A** and **B**. The sum of the vectors is drawn as a dashed line, from the base of the first vector to the tip of the second. As illustrated, the order in which the vectors are connected is not significant as the endpoint is the same graphically whether **A** connects to **B** or **B** connects to **A**. This principle is sometimes called the parallelogram rule.

If more than two vectors are to be combined, additional vectors are simply drawn in accordingly with the sum vector connecting the base of the first to the tip of the final vector.
Subtraction of two vectors can be geometrically defined as follows. To subtract **A** from **B**, place the ends of **A** and **B** at the same point and then draw an arrow from the tip of **A** to the tip of **B**. That arrow represents the vector **B-A**, as illustrated below:

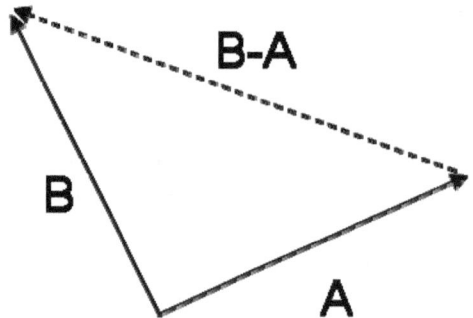

To add two vectors without drawing them, the vectors must be broken down into their orthogonal components using trigonometric functions. Add both x components to get the total x component of the sum vector, then add both y components to get the y component of the sum vector. Use the Pythagorean Theorem and the three trigonometric functions to the get the size and direction of the final vector.

Example: Here is a diagram showing the x and y-components of a vector D1:

Notice that the x-component D1x is adjacent to the angle of 34 degrees.

Thus D1x=36m (cos34) =29.8m

The y-component is opposite to the angle of 34 degrees.

Thus D1y =36m (sin34) = 20.1m

A second vector D2 is broken up into its components in the diagram below using the same techniques. We find that D2y=9.0m and D2x=-18.5m.

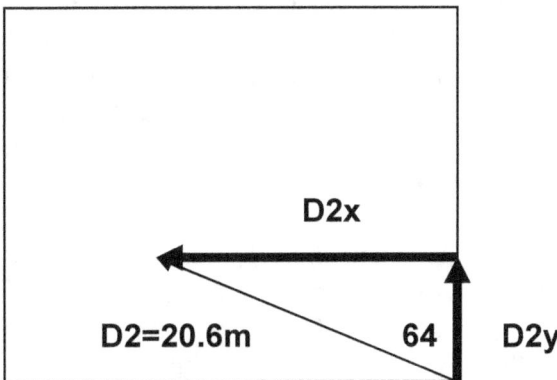

Next we add the x components and the y components to get

DTotal x = 11.3 m   *and*   DTotal y = 29.1 m

Now we have to use the Pythagorean theorem to get the total magnitude of the final vector. And the arctangent function to find the direction. As shown in the diagram below.

DTotal = 31.2 m

tan θ = DTotal y / DTotal x = 29.1m / 11.3 = 2.6   θ = 69 degrees

Skill 5.5 Using fundamental concepts of calculus to model and solve problems

The widespread use of calculus, as applied to problems of physics, can be largely credited to Sir Isaac Newton. Gottfried Wilhelm Leibniz is credited with having simultaneously developed calculus although his work did not have as much physics-oriented character as did Newton's work. The integral and differential, though abstract in and of themselves, can be related to various physical situations and can provide extremely useful analytical tools.

In simple terms, **the derivative of a function is essentially a slope or a rate of change of that function at each point in a defined range**. Thus, the derivative of a function that relates the position of a particle to time is the rate of change of the position, or speed. The derivative of a function over some range is itself a function, and can also be differentiated. Following the previous example, the derivative of the speed of a particle is the rate of change of the speed, or the acceleration. Thus, derivatives can be used to determine additional information about a system based on other information or observations.

In cases where a function involves more than one variable, **partial derivatives** may be involved. These are simply treated as derivatives in terms of the specified variable while assuming that all other variables remain constant. **Partial derivatives are a crucial component of operators such as the gradient, which, in part, determines the direction (in multiple dimensions) of the greatest increase of a function.**

The integral, also aptly named the antiderivative, is the area under the curve of a function in a given range. The function for the speed of a car can be integrated over time to get distance. This is a generalization of the case of constant speed where the speed is simply multiplied by the time traveled to get the distance traveled. The integral performs this multiplication operation over infinitesimally small increments of time thus allowing for calculation in the case of time-varying speed. Integrals may be performed for a single variable even in the case of a function containing several variables. The approach to integration in this case is the same as the approach to partial derivatives: all variables not being integrated are treated as constant.

Vector calculus is often necessary for physical situations where the directions of certain parameters, in addition to the magnitudes, are involved. Disciplines such as classical electrodynamics, for example, must largely be treated with vector calculus. For vector calculus, the choice of coordinate system becomes slightly more complicated, as in some cases the unit vectors are variable. The rectangular coordinate system involves unit vectors that are constant, making it a preferred system of coordinates so long as the expressions are not overly difficult. In these cases it may be beneficial to use other coordinate systems such as cylindrical or spherical.

SUBAREA II   **MECHANICS AND THERMODYNAMICS**

COMPETENCY 6.0   **UNDERSTAND CONCEPTS RELATED TO MOTION IN ONE AND TWO DIMENSIONS, AND APPLY THIS KNOWLEDGE TO SOLVE PROBLEMS THAT REQUIRE THE USE OF ALGEBRA, TRIGONOMETRY, AND GRAPHING**

Skill 6.1   **Applying the terminology, units, and equations used to describe and analyze one and two-dimensional motion**

Kinematics is the part of mechanics that seeks to understand the motion of objects, particularly the relationship between position, velocity, acceleration and time.

[Figure: number line showing $X < 0$, $X = 0$, and $X > 0$ with an object at $X > 0$]

The above figure represents an object and its displacement along one linear dimension.

First we will define the relevant terms:

1. Position or Distance is usually represented by the variable x. It is measured relative to some fixed point or datum called the origin in linear units, meters, for example.

2. Displacement is defined as the change in position or distance which an object has moved and is represented by the variables D, d or Δx. Displacement is a vector with a magnitude and a direction.

3. Velocity is a vector quantity usually denoted with a V or v and defined as the rate of change of position. Typically units are distance/time, m/s for example. Since velocity is a vector, if an object changes the direction in which it is moving it changes its velocity even if the speed (the scalar quantity that is the magnitude of the velocity vector) remains unchanged.

i) **Average velocity:**

The ratio, $\Delta d/\Delta t$ is called the average velocity. Average here denotes that this quantity is defined over a period Δt.

ii) **Instantaneous velocity** is the velocity of an object at a particular moment in time. Conceptually, this can be imagined as the extreme case when Δt is infinitely small.

4. Acceleration represented by a is defined as the rate of change of velocity and the units are m/s^2. Both an average and an instantaneous acceleration can be defined similarly to velocity.

From these definitions we develop the kinematic equations. In the following, subscript i denotes initial and subscript f denotes final values for a time period. Acceleration is assumed to be constant with time.

$$v_f = v_i + at \quad (1)$$

$$d = v_i t + \frac{1}{2}at^2 \quad (2)$$

$$v_f^2 = v_i^2 + 2ad \quad (3)$$

$$d = \left(\frac{v_i + v_f}{2}\right)t \quad (4)$$

In two dimensions the same relationships apply, but each dimension must be treated separately. See **Skill 6.4** for an example problem in two dimensions.

Skill 6.2   **Solving problems involving distance, displacement, speed, velocity, and constant acceleration**

Simple problems involving distance, displacement, speed, velocity, and constant acceleration can be solved by applying the kinematics equations from the proceeding section (**Skill 0006.1**). The following steps should be employed to simplify a problem and apply the proper equations:

1. Create a simple diagram of the physical situation.
2. Ascribe a variable to each piece of information given.
3. List the unknown information in variable form.
4. Write down the relationships between variables in equation form.
5. Substitute known values into the equations and use algebra to solve for the unknowns.
6. Check your answer to ensure that it is reasonable.

Example:

A man in a truck is stopped at a traffic light. When the light turns green, he accelerates at a constant rate of 10 m/s². **a)** How fast is he going when he has gone 100 m? **b)** How fast is he going after 4 seconds? **c)** How far does he travel in 20 seconds?

a=10 m/s²

v$_i$=0 m/s

Solution:
We first construct a diagram of the situation.
In this example, the diagram is very simple, only showing the truck accelerating at the given rate. Next we define variables for the known quantities (these are noted in the diagram):

$$a=10 \text{ m/s}^2; \quad v_i=0 \text{ m/s}$$

Now we will analyze each part of the problem, continuing with the process outlined above.

For part **a)**, we have one additional known variable: d=100 m

The unknowns are: v$_f$ (the velocity after the truck has traveled 100m)

Equation (3) will allow us to solve for v$_f$, using the known variables:

$$v_f^2 = v_i^2 + 2ad$$

$$v_f^2 = (0m/s)^2 + 2(10m/s^2)(100m) = 2000\frac{m^2}{s^2}$$

$$v_f = 45\frac{m}{s}$$

We use this same process to solve part **b)**. We have one additional known variable: t=4 s

The unknowns are: v$_f$ (the velocity after the truck has traveled for 4 seconds)

Thus, we can use equation (1) to solve for v$_f$:

$$v_f = v_i + at$$

$$v_f = 0m/s + (10m/s^2)(4s) = 40m/s$$

For part **c)**, we have one additional known variable: t= 20 s

The unknowns are: d (the distance after the truck has traveled for 20 seconds)

Equation (2) will allow us to solve this problem:

$$d = v_i t + \frac{1}{2}at^2$$

$$d = (0m/s)(20s) + \frac{1}{2}(10m/s^2)(20s)^2 = 2000m$$

Finally, we consider whether these solutions seem physically reasonable. In this simple problem, we can easily say that they do.

Skill 6.3 Interpreting information presented in one or more graphic representations related to distance, displacement, speed, velocity, and constant acceleration

The relationship between time, position or distance, velocity and acceleration can be understood conceptually by looking at a graphical representation of each as a function of time. Simply, the velocity is the slope of the position vs. time graph and the acceleration is the slope of the velocity vs. time graph. If you are familiar with calculus then you know that this relationship can be generalized: velocity is the first derivative and acceleration the second derivative of position.

Here are three examples:

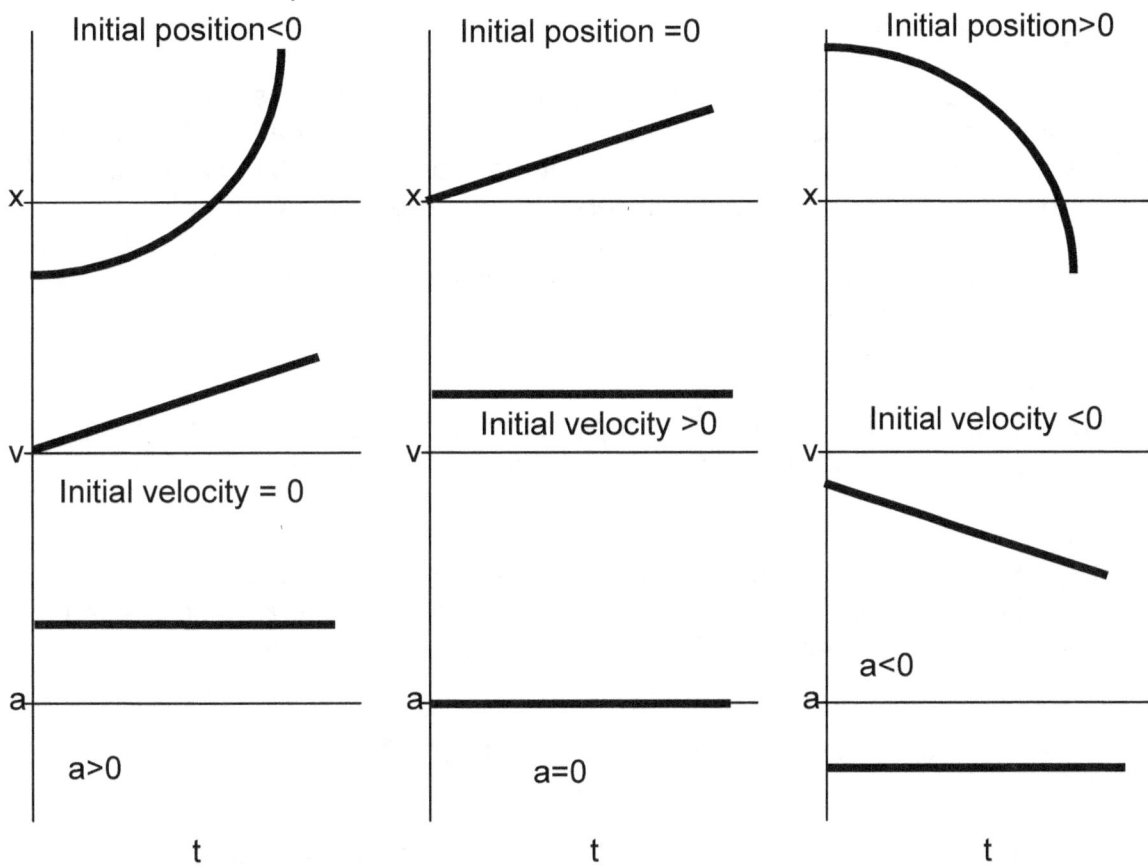

There are three things to notice:
1) In each case acceleration is constant. This isn't always the case, but a simplification for this illustration.
2) A non-zero acceleration produces a position curve that is a parabola.
3) In each case the initial velocity and position are specified separately. The acceleration curve gives the shape of the velocity curve, but not the initial value and the velocity curve gives the shape of the position curve but not the initial position.

TEACHER CERTIFICATION STUDY GUIDE

Skill 6.4 Analyzing the movement of freely falling objects near the surface of the earth

The most common example of an object moving near the surface of the earth is a projectile. A projectile is an object upon which the only force acting is gravity. Some examples:
i) An object dropped from rest
ii) An object thrown vertically upwards at an angle
iii) A canon ball

Once a projectile has been put in motion (say, by a canon or hand) the only force acting on it is gravity which near the surface of the earth is characterized by the acceleration a=g=9.8m/s^2.

This is most easily considered with an example such as the case of a bullet shot horizontally from a standard height at the same moment that a bullet is dropped from exactly the same height. Which will hit the ground first? If we assume wind resistance is negligible, then the acceleration due to gravity is our only acceleration on either bullet and we must conclude that they will hit the ground at the same time. The horizontal motion of the bullet is not affected by the downward acceleration.

Example:
I shoot a projectile at 1000 m/s from a perfectly horizontal barrel exactly 1 m above the ground. How far does it travel before hitting the ground?

Solution:
First figure out how long it takes to hit the ground by analyzing the motion in the vertical direction. In the vertical direction, the initial velocity is zero so we can rearrange kinematic equation 2 from the previous section to give:

$t = \sqrt{\dfrac{2d}{a}}$. Since our displacement is 1 m and a=g=9.8m/s^2, t=0.45 s.

Now use the time to hitting the ground from the previous calculation to calculate how far it will travel horizontally. Here the velocity is 1000m/s and there is no acceleration. So we simple multiply velocity with time to get the distance of 450m.

PHYSICS

TEACHER CERTIFICATION STUDY GUIDE

COMPETENCY 7.0 UNDERSTAND CHARACTERISTICS OF FORCES AND METHODS USED TO MEASURE FORCE, AND SOLVE PROBLEMS INVOLVING FORCES

Skill 7.1 Identifying and analyzing the characteristics of the fundamental forces of nature

There are four fundamental forces that control the behavior of all matter and describe all physical phenomena; gravity, weak interaction, electromagnetism, and strong interaction. Amazingly these same forces control the tiniest sub-nuclear particles, all the life forms on Earth, and the movement of all the planets in the universe. Most physicists believe that we will ultimately be able to understand even these forces as one single, fundamental interaction. Electricity and magnetism were once thought to be separate forces, but we now know that they are parts of a single theory of electromagnetism. Electroweak theory has helped to tie weak interactions with electromagnetism. Likewise, we may eventually combine all four fundamental forces into a single relationship. Note that, while they are called forces, these interactions are not forces in the classic Newtonian sense.

The four forces interact on largely different scales and using different mediators, as will be shown below. They are listed from weakest to strongest; if we assigned gravity a relative strength of 1, then the weak interaction would be 10^{25} times stronger, the electromagnetic by a factor of 10^{36}, and the strong interaction by a factor of 10^{38}.

Gravity: Though gravity is the weakest interaction, it has infinite range and affects all masses. Thus, it is responsible for the structure of galaxies, the orbits of planets, and the expansion of the universe as well as everyday phenomena such as the falling of objects. Gravity is mediated by elementary particles known as gravitons and is currently understood through the theory of General Relativity.

Weak interaction: Despite the name, weak interactions are much stronger than gravity but with a much smaller range. They are responsible for certain phenomena occurring in the atomic nucleus, including beta decay, and other occurrences on this size scale. Weak interactions are mediated by bosons and understood through the Electroweak Theory.

Electromagnetism: This force acts on charged particles and, like gravity, has a large enough range that we can observe its effect in everyday occurrences. Electromagnetism is responsible for a wide range of phenomena ranging from rainbows and sound to lasers and the structure of metal. It is mediate by photons and understood through the theory of Quantum Electrodynamics (QED).

PHYSICS

Strong interactions: Strong interactions work primarily on sub-atomic particles though their range is theoretically infinite. They are responsible for holding protons and neutrons together. Nuclear force, which holds the nucleus together, is a byproduct of strong interactions. This demonstrates the large strength of these interactions, since electromagnetic forces would otherwise cause the protons in the nucleus to repel one another. Strong interactions are mediated by gluons and understood through the theory of Quantum Chromodynamics.

Skill 7.2   Identifying forces acting in a given situation

Some of the common forces that act on a body are the following:

Gravity
This is the force that pulls a body towards the center of the earth, i.e. downwards, and is also called the weight of the body. It is given by
$$W = mg$$
Where m is the mass of the body and $g=9.81$ m/s² is the acceleration due to gravity.

Normal force
When a body is pressed against a surface it experiences a reaction force that is perpendicular to the surface and in the direction away from the surface. For instance, an object resting on a table experiences an upward reaction force from the table that is equal and opposite to the force that the object exerts on the table. When the table is horizontal and no additional force is being applied to the object, the normal force is equal to the weight of the object.

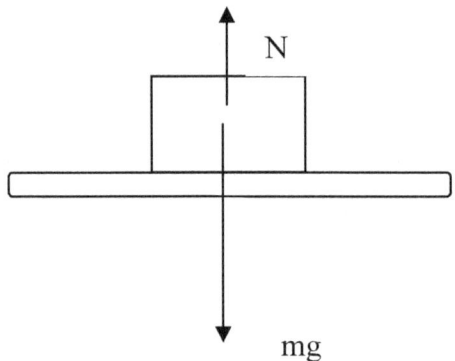

Friction
Friction is the force on a body that opposes its sliding over a surface. This force is due to the bonding between the two surfaces and is greater for rough surfaces. It acts in the direction opposite to the force attempting to move the object. When the object is at rest, the frictional force is known as **static friction**. The frictional force on an object in motion is known as **kinetic friction**.

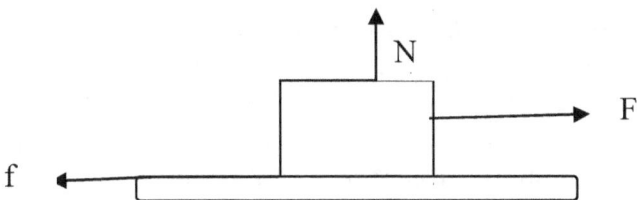

The frictional force is usually directly proportional to the normal force and can be calculated as $F_f = \mu F_n$ where μ is either the coefficient of static friction or kinetic friction depending on whether the object is at rest or in motion.

Tension/Compression

Tension is the force that acts in a rope, cable or rod that is attached to something and is being pulled. Tension acts along the cord. When a hand pulls a rope attached to a box, for instance, the tension T in the rope acts to pull the rope apart while it works on the box and the hand in the opposite direction as shown below:

Compression is the opposite of tension in that the force acts to shorten a rigid body instead of pulling it apart.

Pressure

For a description of fluid pressure and the force of buoyancy see **Skill 1.1c**.

Net force

The forces that act on a body come from many different sources. Their effect on a body, however, is the same; a change in the state of motion of the body as given by Newton's laws of motion. Therefore, once we identify the magnitude and direction of each force acting on a body, we can combine the effect of all the forces together using vector addition and find the net force.

Skill 7.3   Analyzing experimental designs for measuring forces

Designing an experiment to accurately measure a force directly can be extremely challenging and indirect methods are often required. For example, to measure the force on a moving object, it may be simpler to measure the position of the object with time, differentiate the results twice to get the acceleration and then multiply by the mass of the object in accordance with $F = ma$. Thus, force measurement experiments may often be more complicated than simply using a mechanical scale delineated in Newtons.

Indirect force measurement through fields or potentials

Forces may be measured indirectly through a field or potential measurement. Since it is impossible to attach a tiny spring scale to an electron in order to measure the force of an electric field upon it, a measurement of the electric field or electric potential can be used instead. Such measurements may be taken by way of a field meter or multimeter, providing information about the field or potential from which the force on an electron (or other object) can be calculated.

Indirect force measurements through material characteristics

Properties of various materials can also be used to indirectly measure a force. The piezoelectric effect is one example that results from the ability of certain materials, such as some crystals and ceramics, to produce an electric potential when they are stressed mechanically. Thus, these natural "pressure sensors" can be used for force measurements through measurement of the voltage produced in a given situation. One particular application of piezoelectric materials is in equipment for fine weight measurements that can be used to measure the gravitational force on small or low-mass objects.

The temperature and volume of a confined gas may also be used to measure force by using, for example, a piston. Force applied to the piston causes an increase in the pressure of the gas and, by the ideal gas law ($PV = nRT$), a corresponding change in the volume and temperature of the gas. The pressure can be calculated from these parameters (or measured directly), thus allowing the calculation of the force on the piston.

Stages of experimental design

Experimental apparatuses for measurement of forces, therefore, involve several aspects. First, the transducer, which may be more or less a direct measurement of the force, acts as a converter of force into some intelligible signal. In some cases, as with the examples of the piston or the piezoelectric potential, the magnitude of the force must be calculated from the signal (whether electric or otherwise) through a theoretical relationship determined by the characteristics of the material or materials used. Second, for the more indirect force measurements, computer equipment may be required for calculating the force based on other parameters in the experiment. In the example of the moving object, it is most convenient to use a computer for performing the differentials, although, for sufficiently small data sets, this can be done "by hand." Regardless of the complexity, an algorithm for calculating the force values must be followed, and this is as much part of the experimental apparatus as the transducer equipment. In simple cases, however, such as spring scales, the force can be measured by simply looking at the reading.

Skill 7.4 Applying graphic solutions to solve problems involving the vector nature of force

Vectors are used in representing any quantity that has both magnitude and direction. We must know not only that the force of gravity has a magnitude of 9.8 m/s^2 times the mass of an object, but that it is directed toward the center of the earth.

When we wish to analyze a physical situation involving vector quantities such as force, the first step is typically the creation of a diagram. Free body diagrams are simple sketches that show all the objects and forces in a given physical situation. This makes them very useful for understanding and solving physical problems. The objects involved are drawn and arrows are used to represent the vector quantities which are labeled appropriately. In precise diagrams drawn to scale on graph paper, the magnitude of each force is indicated by the length of the arrow and the exact angles between the vectors will be depicted. Otherwise, angles and magnitudes may simply be noted on the diagram.

Below is an example of a free body diagram. See the next section for an example where the net force is calculated algebraically. **For graphical and trigonometric addition of all kinds of vectors, including forces, see Skill 5.4.**

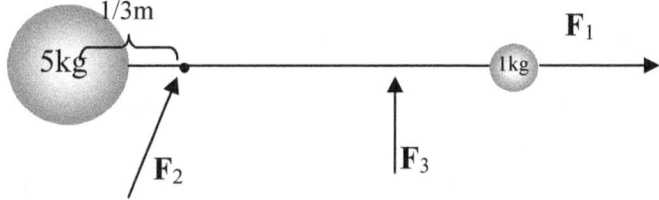

TEACHER CERTIFICATION STUDY GUIDE

Skill 7.5 Determining the resultant of two or more forces algebraically

Since forces are vectors, several forces acting on any one object may be combined using the same graphical or trigonometric/algebraic methods used for other vectors (see **Skill 5.4**). An example problem is given below.

Problem:
Find the net force on a 5 Kg box sliding down an inclined surface at an angle of 30^0 with the horizontal if the coefficient of friction of the surface is 0.5.

Solution:
There are three forces acting on the box, gravity, the normal force and the frictional force. We can resolve these forces along the inclined plane and perpendicular to the plane and find the net force in each direction.

Perpendicular to the plane:
The component of the gravitational force perpendicular to the plane = mgcos30 = 5 x 9.8 x 0.87 = 42.6N
The normal force acting on the box is equal and opposite to the perpendicular component of the gravitational force. Thus the net force on the box perpendicular to the plane is zero.

Along the inclined plane:

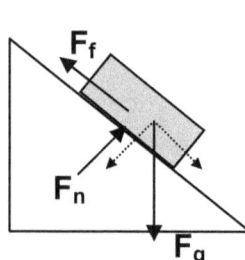

The component of the gravitational force down the plane =mgsin30 = 5 x 9.8 x 0.5 = 24.5N
The force of friction up the plane = μ **F**$_n$ = 0.5 x 42.6 = 21.3N
Thus the net force on the box acts down the plane and is equal to 24.5 – 21.3 = 3.2N

Skill 7.6 Applying the concepts of force, pressure, and density

Pressure is the force applied per unit area on a surface and is measured in the units N/m^2 in the SI system. Density is the mass per unit volume and is measured in the SI units kg/m^3. Pressure and density (in place of force and mass) are useful concepts when we are dealing with an extended mass of substance such as a fluid as opposed to a lump of matter such as a solid ball.

The weight of a column of fluid creates hydrostatic pressure. Common situations in which we might analyze hydrostatic pressure include tanks of fluid, a swimming pool, or the ocean. Also, **atmospheric pressure** is an example of hydrostatic pressure. Because hydrostatic pressure results from the force of gravity interacting with the mass of the fluid or gas, for an incompressible fluid it is governed by the following equation:

$$P = \rho g h$$

where P=hydrostatic pressure
ρ=density of the fluid
g=acceleration of gravity
h=height of the fluid column

Example: How much pressure is exerted by the water at the bottom of a 5 meter swimming pool filled with water?

Solution: We simply use the equation from above, recalling that the acceleration due to gravity is 9.8m/s² and the density of water is 1000 kg/m³.

$$P = \rho g h = 1000 \frac{kg}{m^3} \times 9.8 \frac{m}{s^2} \times 5m = 49,000\,Pa = 49kPa$$

Skill 7.7 Applying Pascal's principle to analyze fluid statics

According to Pascal's principle, when pressure is applied to an enclosed fluid, it is transmitted undiminished to all parts of the fluid. For instance, if an additional pressure P_0 is applied to the top surface of a column of liquid of height h as described above, the pressure at the bottom of the liquid will increase by P_0 and will be given by $P = P_0 + \rho g h$.

This principle is used in devices such as a hydraulic lift (shown below) which consists of two fluid-filled cylinders, one narrow and one wide, connected at the bottom. Pressure P (force = P X A1) applied on the surface of the fluid in the narrow cylinder is transmitted undiminished to the wider cylinder resulting in a larger net force (P X A2) transmitted through its surface. Thus a relatively small force is used to lift a heavy object. This does not violate the conservation of energy since the small force has to be applied through a large distance to move the heavy object a small distance.

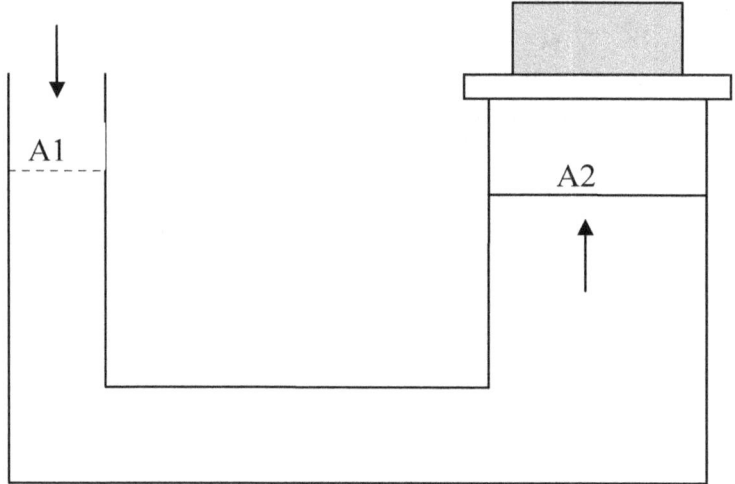

Skill 7.8   Applying Archimedes' principle to problems involving buoyancy and flotation

Archimedes discovered through observation that, for an object in a fluid, "the upthrust is equal to the weight of the displaced fluid" and the weight of displaced fluid is directly proportional to the volume of displaced fluid. The second part of his discovery is useful when we want to, for instance, determine the volume of an oddly shaped object. We determine its volume by immersing it in a graduated cylinder and measuring how much fluid is displaced. We explore his first observation in more depth below.

Today, we call Archimedes' "upthrust" **buoyancy**. Buoyancy is the force produced by a fluid on a fully or partially immersed object. The buoyant force ($F_{buoyant}$) is found using the following equation:

$$F_{buoyant} = \rho V g$$

> where ρ=density of the fluid
> V=volume of fluid displaced by the submerged object
> g=the acceleration of gravity

Notice that the buoyant force opposes the force of gravity. For an immersed object, a gravitational force (equal to the object's mass times the acceleration of gravity) pulls it downward, while a buoyant force (equal to the weight of the displaced fluid) pushes it upward.

Also note that, from this principle, we can predict *whether* an object will sink or float in a given liquid. We can simply compare the density of the material from which the object is made to that of the liquid. If the material has a lower density, it will float; if it has a higher density it will sink. Finally, if an object has a density equal to that of the liquid, it will neither sink nor float.

Example: Will gold (ρ=19.3 g/cm^3) float in water?

Solution: We must compare the density of gold with that of water, which is 1 g/cm^3.

$$\rho_{gold} > \rho_{water}$$

So, gold will sink in water.

Example: Imagine a 1 m³ cube of oak (530 kg/m³) floating in water. What is the buoyant force on the cube and how far up the sides of the cube will the water be?

Solution: Since the cube is floating, it has displaced enough water so that the buoyant force is equal to the force of gravity. Thus the buoyant force on the cube is equal to its weight 1X530X9.8 N = 5194 N.

To determine where the cube sits in the water, we simply the find the ratio of the wood's density to that of the water:

$$\frac{\rho_{oak}}{\rho_{water}} = \frac{530\,kg/m^3}{1000\,kg/m^3} = 0.53$$

Thus, 53% of the cube will be submerged. Since the edges of the cube must be 1m each, the top 0.47m of the cube will appear above the water.

COMPETENCY 8.0 UNDERSTAND AND APPLY THE LAWS OF MOTION (INCLUDING RELATIVITY) AND CONSERVATION OF MOMENTUM

Skill 8.1 Analyzing the characteristics of each of Newton's laws of motion and giving examples of each

Newton's first law of motion: "An object at rest tends to stay at rest and an object in motion tends to stay in motion with the same speed and in the same direction unless acted upon by an unbalanced force". Prior to Newton's formulation of this law, being at rest was considered the natural state of all objects, because at the earth's surface we have the force of gravity working at all times which causes nearly any object put into motion to eventually come to rest. Newton's brilliant leap was to recognize that an unbalanced force changes the motion of a body, whether that body begins at rest or at some non-zero speed.

We experience the consequences of this law everyday. For instance, the first law is why seat belts are necessary to prevent injuries. When a car stops suddenly, say by hitting a road barrier, the driver continues on forward until acted upon by a force. The seat belt provides that force and distributes the load across the whole body rather than allowing the driver to fly forward and experience the force against the steering wheel.

Example: A skateboarder is riding her skateboard down a road. The skateboard has a constant speed of 5 m/s. Then the skateboard hits a rock and stops suddenly. Since the rider has nothing to stop her when the skateboard stops, she will continue to travel at 5 m/s until she hits the ground.

Newton's second law of motion: "The acceleration of an object as produced by a net force is directly proportional to the magnitude of the net force, in the same direction as the net force, and inversely proportional to the mass of the object". In the equation form, it is stated as $F = ma$, force equals mass times acceleration. It is important to remember that this is the net force and that forces are vector quantities. Thus if an object is acted upon by 12 forces that sum to zero, there is no acceleration. Also, this law embodies the idea of inertia as a consequence of mass. For a given force, the resulting acceleration is proportionally smaller for a more massive object because the larger object has more inertia.

Example:
A ball is dropped from a building. The mass of the ball is 2 kg. The acceleration of the object is 9.8 m/s² (gravitational acceleration). Therefore, the force acting on the ball is
$F = ma \Rightarrow F = 2 \text{ kg} \times 9.8 \text{ m/s}^2 \Rightarrow F = 19.6 \text{ N}$

Newton's third law of motion: "For every action, there is an equal and opposite reaction". This statement means that in every interaction, there is a pair of forces acting on the two interacting objects. The size of the force on the first object equals the size of the force on the second object. The direction of the force on the first object is opposite to the direction of the force on the second object.

Example: A box is sitting on a table. The mass of the box is 4 kg. Because of the effects of gravity, the box is applying a force of 39.2 N on the table. The table does not break or shift under the force of the box. This implies that the table is applying a force of 39.2 N on the box. Note that the force that the table is applying to the box is in the opposite direction to the force that the box is applying to the table.

Here are a few more examples:
1. The propulsion/movement of fish through water: A fish uses its fins to push water backwards. The water pushes back on the fish. Because the force on the fish is unbalanced the fish moves forward.
2. The motion of car: A car's wheels push against the road and the road pushes back. Since the force of the road on the car is unbalanced the car moves forward.
3. Walking: When one pushes backwards on the foot with the muscles of the leg, the floor pushes back on the foot. If the forces of the leg on the foot and the floor on the foot are balanced, the foot will not move and the muscles of the body can move the other leg forward.

Skill 8.2   Applying Newton's laws of motion in solving problems

Newton's laws of motion can be used together or separately to analyze a variety of physical situations. Simple examples are provided below and the importance of each law is highlighted.

Problem:
A 10 kg object moves across a frictionless surface at a constant velocity of 5 m/s. How much force is necessary to maintain this speed?

Solution:
Both Newton's first and second laws can help us understand this problem. First, the first law tells us that this object will continue its state of uniform speed in a straight line (since there is no force acting upon it). Additionally, the second law tells that because there is no acceleration (velocity is constant), no force is required. Thus, zero force is necessary to maintain the speed of 5 m/s.

Problem:
A car is driving down a road at a constant speed. The mass of the car is 400 kg. The force acting on the car is 4000 N and the force is in the same direction as the acceleration. What is the acceleration of the car?

Solution:

$$F = ma \Rightarrow a = \frac{F}{m} \Rightarrow a = \frac{4000\text{N}}{400 kg} \Rightarrow a = 10 m/s^2$$

Problem:
For the arrangement shown, find the force necessary to overcome the 500 N force pushing to the left and move the truck to the right with an acceleration of 5 m/s^2.

Solution:
The net force on the truck acting towards the right is F − 500N.
Using Newton's second law, F−500N = 1000kg × 5 m/s^2.
Solving for F, we get F = 5500 N.

Problem:
An astronaut with a mass of 95 kg stands on a space station with a mass of 20,000 kg. If the astronaut is exerting 40 N of force on the space station, what is the acceleration of the space station and the astronaut?

Solution
To find the acceleration of the space station, we can simply apply Newton's second law:

$$A_s = \frac{F}{m_s} = \frac{40N}{20000kg} = 0.002 \, m/s^2$$

To find the acceleration of the astronaut, we must first apply Newton's third law to determine that the space station exerts an opposite force of -40 N on the astronaut. Here the minus sign simply denotes that the force is directed in the opposite direction. We can then calculate the acceleration, again using Newton's second law:

$$A_a = \frac{F}{m_a} = \frac{-40N}{95kg} = -0.42 \, m/s^2$$

Skill 8.3   Solving problems involving gravitational and frictional forces

Newton's universal law of gravitation states that any two objects experience a force between them as the result of their masses. Specifically, the force between two masses m_1 and m_2 can be summarized as

$$F = G\frac{m_1 m_2}{r^2}$$

where G is the gravitational constant ($G = 6.672 \times 10^{-11} \, Nm^2/kg^2$), and r is the distance between the two objects.

The weight of an object is the result of the gravitational force of the earth acting on its mass. The acceleration due to Earth's gravity on an object is 9.81 m/s². Since force equals mass * acceleration, the magnitude of the gravitational force created by the earth on an object is

$$F_{Gravity} = m_{object} \cdot 9.81 \, m/s^2$$

Important things to remember:

1. The gravitational force is proportional to the masses of the two objects, but *inversely* proportional to the *square of the distance* between the two objects.
2. When calculating the effects of the acceleration due to gravity for an object above the earth's surface, the distance above the surface is ignored because it is inconsequential compared to the radius of the earth. The constant figure of 9.81 m/s² is used instead.

Problem: Two identical 4 kg balls are floating in space, 2 meters apart. What is the magnitude of the gravitational force they exert on each other?

Solution:

$$F = G\frac{m_1 m_2}{r^2} = G\frac{4 \times 4}{2^2} = 4G = 2.67 \times 10^{-10} \, N$$

In the real world, whenever an object moves its motion is opposed by a force known as friction. How strong the frictional force is depends on numerous factors such as the roughness of the surfaces (for two objects sliding against each other) or the viscosity of the liquid an object is moving through. Most problems involving the effect of friction on motion deal with sliding friction. This is the type of friction that makes it harder to push a box across cement than across a marble floor.

When you try and push an object from rest, you must overcome the maximum **static friction** force to get it to move. Once the object is in motion, you are working against **kinetic friction** which is smaller than the static friction force previously mentioned. Sliding friction is primarily dependent on two things, the **coefficient of friction (μ)** which is primarily dependent on roughness of the surfaces involved and the amount of force pushing the two surfaces together. This force is also known as the **normal force (F_n)**, the perpendicular force between two surfaces. When an object is resting on a flat surface, the normal force is pushing opposite to the gravitational force – straight up. When the object is resting on an incline, the normal force is less (because it is only opposing that portion of the gravitational force acting perpendicularly to the object) and its direction is perpendicular to the surface of incline but at an angle from the ground. Therefore, for an object experiencing no external action, the magnitude of the normal force is either equal to or less than the magnitude of the gravitational force **(F_g)** acting on it. The frictional force **(F_f)** acts perpendicularly to the normal force, opposing the direction of the object's motion.

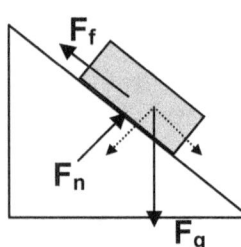
The frictional force is normally directly proportional to the normal force and, unless you are told otherwise, can be calculated as $F_f = \mu F_n$ where μ is either the coefficient of static friction or kinetic friction depending on whether the object starts at rest or in motion. In the first case, the problem is often stated as "how much force does it take to start an object moving" and the frictional force is given by $F_f > \mu_s F_n$ where μ_s is the coefficient of static friction. When questions are of the form "what is the magnitude of the frictional force opposing the motion of this object," the frictional force is given by $F_f = \mu_k F_n$ where μ_k is the coefficient of kinetic friction.

There are several important things to remember when solving problems about friction.
1. The frictional force acts in opposition to the direction of motion.
2. The frictional force is proportional, and acts perpendicular to, the normal force.
3. The normal force is perpendicular to the surface the object is lying on. If there is a force pushing the object against the surface, it will increase the normal force.

Problem:
A woman is pushing an 800N box across the floor. She pushes with a force of 1000 N. The coefficient of kinetic friction is 0.50. If the box is already moving, what is the force of friction acting on the box?

Solution:
First it is necessary to solve for the normal force.
$F_n = 800N + 1000N (\sin 30°) = 1300N$
Then, since $F_f = \mu F_n = 0.5*1300 = 650N$

Skill 8.4   Solving problems involving springs and force constants

The above diagram is an example of a **Hookean system** (a spring, wire, rod, etc.) where the spring returns to its original configuration after being displaced and then released. When the spring is stretched a distance x, the restoring force exerted by the spring is expressed by Hooke's Law

$$F = -kx$$

The minus sign indicates that the restoring force is always opposite in direction to the displacement. The variable k is the spring constant and measures the stiffness of the spring in N/m.

The period of simple harmonic motion for a Hookean spring system is dependent upon the mass (m) of the spring and the stiffness of the spring (k) and is given by

$$T = \frac{1}{f} = \frac{1}{\frac{\omega}{2\pi}} = \frac{2\pi}{\omega} = \frac{2\pi}{\sqrt{\frac{k}{m}}} = 2\pi\sqrt{\frac{m}{k}}$$

Problem:
Each spring in the above diagram has a stiffness of $k = 20 N/m$. The mass of the object connected to the spring is 2 kg. Ignoring friction forces, find the period of motion.

Solution:
Utilizing Hooke's Law, the net restoring force on the spring would be

$$F = -(20 N/m)x - (20 N/m)x = -(40 N/m)x$$

Comparison with $F = -kx$ shows the equivalent k to be 40 N/m.
Using the above formula for our problem, we have

$$T = 2\pi\sqrt{\frac{m}{k}} = 2\pi\sqrt{\frac{2 kg}{40 N/m}} = 2(3.14)\sqrt{0.05} = 1.4 s$$

Skill 8.5 Applying the conservation of momentum to analyze and solve problems

The law of conservation of momentum states that the total momentum of an *isolated system* (not affected by external forces and not having internal dissipative forces) always remains the same. For instance, in any collision between two objects in an isolated system, the total momentum of the two objects after the collision will be the same as the total momentum of the two objects before the collision. In other words, any momentum lost by one of the objects is gained by the other. In one dimension this is easy to visualize.

Imagine two carts rolling towards each other as in the diagram below

Before the collision, cart **A** has 250 kg m/s of momentum, and cart **B** has −600 kg m/s of momentum. In other words, the system has a total momentum of −350 kg m/s of momentum.

After the collision, the two cards stick to each other, and continue moving. How do we determine how fast, and in what direction, they go?

We know that the new mass of the cart is 80kg, and that the total momentum of the system is −350 kg m/s. Therefore, the velocity of the two carts stuck together must be $\dfrac{-350}{80} = -4.375\, m/s$

Conservation of momentum works the same way in two dimensions, the only change is that you need to use vector math to determine the total momentum and any changes, instead of simple addition.

Imagine a pool table like the one below. Both balls are 0.5 kg in mass.

Before the collision, the white ball is moving with the velocity indicated by the solid line and the black ball is at rest.
After the collision the black ball is moving with the velocity indicated by the dashed line (a 135° angle from the direction of the white ball).

With what speed, and in what direction, is the white ball moving after the collision?

$p_{white/before} = .5 \cdot (0,3) = (0,1.5) \quad p_{black/before} = 0 \quad p_{total/before} = (0,1.5)$

$p_{black/after} = .5 \cdot (2\cos 45, 2\sin 45) = (0.71, 0.71)$

$p_{white/after} = (-0.71, 0.79)$

i.e. the white ball has a velocity of
$v = \sqrt{(-.71)^2 + (0.79)^2} = 1.06 \, m/s$
and is moving at an angle of
$\theta = \tan^{-1}\left(\dfrac{0.79}{-0.71}\right) = -48°$ from the horizontal

Skill 8.6 Understanding the implications of special relativity for the laws of motion

Postulates of special relativity
Einstein's theory of special relativity built upon the foundation of Galilean relativity and incorporated an analysis of electromagnetics. Galilean relativity is based upon the concept that the laws of physics are invariant with respect to inertial frames of reference. In the context of classical electrodynamics, it is found that Maxwell's equations do not imply any variation in the speed of light with respect to the relative motion of the source and observer. As a result, the second fundamental postulate of special relativity is a statement of the invariance of the speed of light, c, with respect to inertial reference frames.

The postulates of special relativity in brief:
1. Special principle of relativity: The laws of physics are same in all inertial frames of reference.
2. Invariance of c: The speed of light in a vacuum is a universal constant for all observers, regardless of the inertial frame of reference or the relative motion of the source.

Force and acceleration
One of the implications of these postulates is that the relativistic mass m of a particle is dependent upon its velocity v. This relationship between the rest mass m_0 and the relativistic mass is expressed through the Lorentz factor γ.

$$m = \gamma m_0 = \frac{m_0}{\sqrt{1 - \frac{v^2}{c^2}}}$$

This has a particular effect on the relationship of force F and acceleration a, which is generally expressed using the momentum p.

$$F = \frac{\partial p}{\partial t} = \frac{\partial (mv)}{\partial t}$$

In the context of special relativity, where the mass is dependent on the velocity (which is a function of time), the above equation does not simplify to F = ma, as it does in Newtonian mechanics.

$$F = v \frac{\partial m}{\partial t} + ma = m_0 \left(v \frac{\partial \gamma}{\partial t} + \gamma a \right)$$

Since the Lorentz factor increases asymptotically towards infinity as the speed of the particle or object approaches the speed of light, the force required to accelerate the particle also approaches infinity. It would then require infinite energy to accelerate an object to the speed of light, thus making c the "speed limit of the universe." In cases where the speed of the object is significantly less than c, the Lorentz factor is close to unity, thus making Newtonian mechanics an accurate approximation. (This concept is similar to the correspondence principle of quantum mechanics; in this case, relativistic mechanics becomes Newtonian mechanics in the limit as v goes to zero.)

Velocity

As mentioned, the postulates of special relativity seem to imply a universal speed limit. Thus, if this is to be the case in all inertial frames of reference, the observed speed of an object cannot be greater than the speed of light regardless of the velocities of the object and inertial frame of reference as measured in any other inertial frame of reference. As a result, although a particular frame of reference may be moving in one direction at speed w close to light speed and a particular object is moving in the opposite direction at a speed v, likewise near c, the speed of the object as measured from the reference frame *cannot* simply be the sum of v and w. Instead, special relativity uses a different approach to the calculation resulting in a relative velocity given by the following formula.

$$v' = \frac{v + w}{1 + vw/c^2}$$

Conservation of energy

In classical mechanics the behavior of particles or objects can be determined through the application of the conservation of energy and the conservation of momentum. Although, in this case, these two are seemingly independent concepts, in the case of relativistic mechanics they are shown to be interdependent. This interdependence results from the relationship of mass and energy that is derived from the application of the postulates of special relativity.

$$E^2 = (mc^2)^2 + (pc)^2$$

Thus, mass and energy (and, therefore, energy and momentum) are interdependent characteristics of the particle or object. The conservation laws for classical momentum and energy are then joined into a more general expression of the conservation of energy.

TEACHER CERTIFICATION STUDY GUIDE

COMPETENCY 9.0 UNDERSTAND AND APPLY THE PRINCIPLE OF CONSERVATION OF ENERGY AND THE CONCEPTS OF ENERGY, WORK, AND POWER

Skill 9.1 Analyzing mechanical systems in terms of work, power, and conservation of energy

In physics, work is defined as force times distance $W = F \cdot s$. Work is a scalar quantity, it does not have direction, and it is usually measured in Joules ($N \cdot m$). It is important to remember, when doing calculations about work, that the only part of the force that contributes to the work is the part that acts in the direction of the displacement. Therefore, sometimes the equation is written as $W = F \cdot s \cos\theta$, where θ is the angle between the force and the displacement.

Problem:
A man uses 6N of force to pull a 10kg block, as shown below, over a distance of 3 m. How much work did he do?

Solution:

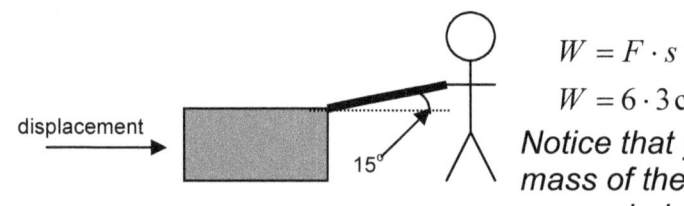

$W = F \cdot s \cos\theta$

$W = 6 \cdot 3 \cos 15 = 17.4 J$

Notice that you did not actually need the mass of the box in order to calculate the work done.

Energy is also defined, in relation to work, as the ability of an object to do work. As such, it is measured in the same units as work, usually Joules. Most problems relating work to energy are looking at two specific kinds of energy. The first, kinetic energy, is the energy of motion. The heavier an object is and the faster it is going, the more energy it has resulting in a greater capacity for work. The equation for kinetic energy is: $KE = \frac{1}{2}mv^2$.

Problem:
A 1500 kg car is moving at 60m/s down a highway when it crashes into a 3000kg truck. In the moment before impact, how much kinetic energy does the car have?

Solution:
$KE = \frac{1}{2}mv^2 = \frac{1}{2} \cdot 1500 \cdot 60^2 = 2.7 \times 10^6 J$

The other form of energy frequently discussed in relationship to work is gravitational potential energy, or potential energy, the energy of position.

PHYSICS

Potential energy is calculated as $PE = mgh$ where h is the distance the object is capable of falling.

Problem:
Which has more potential energy, a 2 kg box held 5 m above the ground or a 10 kg box held 1 m above the ground?

Solution:

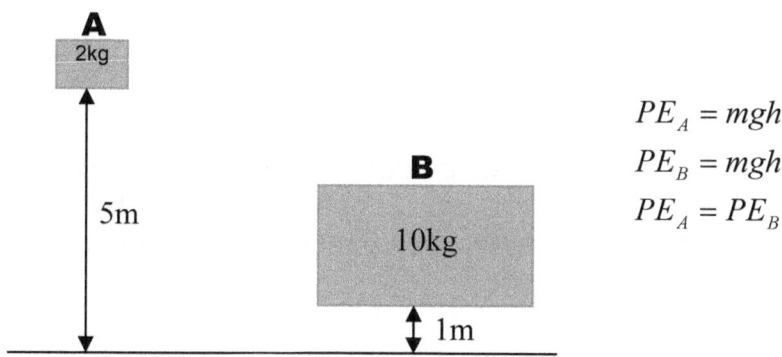

$$PE_A = mgh = 2 \cdot g \cdot 5 = 10g$$
$$PE_B = mgh = 10 \cdot g \cdot 1 = 10g$$
$$PE_A = PE_B$$

The power expended by a system can be defined as either the rate at which work is done by the system or the rate at which energy is transferred from it. There are many different measurements for power, but the one most commonly seen in physics problems is the Watt which is measured in Joules per second. Another commonly discussed unit of power is horsepower, and 1hp=746 W.

The **average power** of a system is defined as the work done by the system divided by the total change in time:

$\overline{P} = \dfrac{W}{\Delta t}$ ⇒ Where $\overline{P}$ = average power, W = work and Δt = change in time

The average power can also be written in terms of energy transfer ⇒ $\overline{P} = \dfrac{\Delta E}{\Delta t}$ and used the same way that the equation for work is used.

Problem:
A woman standing in her 4th story apartment raises a 10kg box of groceries from the ground using a rope. She is pulling at a constant rate, and it takes her 5 seconds to raise the box one meter. How much power is she using to raise the box?

Solution:

$$P = W/t$$

$$P = \frac{F \cdot s}{t} = \frac{mgh}{t} = \frac{10*9.8*1}{5} = 19.6W$$

Notice that because she is pulling at a constant rate, you don't need to know the actual distance she has raised the box. 2 meters in 10 seconds would give you the same result as 5 meters in 25 seconds.

Instantaneous power is the power measured or calculated in an instant of time. Since instantaneous power is the rate of work done when Δt is approaching 0s, the power is then written in derivate form:

$P = \frac{dW}{dt}$ ⇒ Where P = average power, dW = work and dt = change in time.

Since $W = Fs\cos\phi$, for a constant force the above equation can be written as:

$$P = \frac{dW}{dt} \Rightarrow P = \frac{d(Fs\cos\phi)}{dt} \Rightarrow P = \frac{(F\cos\phi)ds}{dt} \Rightarrow P = F\cos\phi\left(\frac{ds}{dt}\right) \Rightarrow P = Fv\cos\phi$$

where v is the velocity of the object.

Skill 9.2 Using the concept of conservation of energy to solve problems

According to the concept of conservation of energy, the energy in an isolated system remains the same although it may change in form. For instance, potential energy can become kinetic energy and kinetic energy, depending on the system, can become thermal or heat energy. Solving energy conservation problems depends on knowing the types of energy one is dealing with in a particular situation and assuming that the sum of all the different types of energy remains constant. Below we will discuss several different examples.

Example:
A rollercoaster at the top of a hill has a certain potential energy that will allow it to travel down the track at a speed based on its potential energy and friction with the track itself. At the bottom of the hill, when it has reached a stop, its potential energy is zero and all of the energy has been transferred from potential energy to kinetic energy (movement) and thermal energy (heat derived from friction). The equation below describes the relationship between potential energy and other forms of energy in this case:

Potential Energy = Kinetic energy(movement) + heat energy(friction)

Problem:
A skier travels down a ski slope with negligible friction. He begins at 100 meters in height, drops to a much lower level and ends at 90 meters in height. What is the skier's velocity at the 90 meter height?

100m   90m

The skier's initial energy is only potential energy and is given by mgh = 100mg where m is the mass of the skier.
The skier's final energy is the sum of his potential and kinetic energies and is given by mgh + 1/2mv² = 90mg + 1/2mv² where v is the skier's velocity.
Using the principle of conservation of energy we know that the initial and final energies must be equal. Hence
$$100mg = 90mg + 1/2mv^2$$

Thus $1/2mv^2$ = 10mg; $1/2v^2$ = 10g; v = 14m/s since (g=9.81m/s²)

Problem:
d. A pebble weighing 10 grams is placed in a mass-less frictionless sling shot, with a spring constant of 200N/m, that is stretched back 0.5 meters. What is the total energy of the system before the pebble is released? What is the final height of the pebble if it is shot straight up and the effects of air resistance are negligible?

Solution:
If the initial height of the pebble is h = 0, total energy is given by
E = ½ mv² + mgh + ½ kx² = 0 + 0 + 0.5(200)(0.5)² = 25 Joules

At its final height, the velocity of the pebble will be zero. Since
E = ½ mv² + mgh + ½ kx², from the principle of conservation of energy
25 Joules = 0 + 0.010kg (9.81)h + 0 and h = 255 m

Skill 9.3   Solving problems using the work-energy theorem

The work-energy theorem states that **the amount of work done on an object is equal to its change in mechanical energy (kinetic or potential energy)**. Specifically, in systems where multiple forces are at work, the energy change of the system is the work done by the *net* force on the object. Problems dealing with the work-energy theorem may look at changes in kinetic energy, changes in potential energy, or some combination of the two. It is also important to remember that only external forces can cause changes in an object's total amount of mechanical energy. Internal forces, such as spring force or gravity, only lead to conversions between kinetic and potential energy rather than changes in the total level of mechanical energy.

Problem:
A woman driving a 2000 kg car along a level road at 30 m/s takes her foot off the gas to see how far her car will roll before it slows to a stop. She discovers that it takes 150m. What is the average force of friction acting on the car?

Solution:

$$W = \Delta KE$$

$$f \cdot s \cos\theta = \tfrac{1}{2} m v_{final}^2 - \tfrac{1}{2} m v_{initial}^2$$

$$f \cdot 150 \cdot (-1) = \tfrac{1}{2} \cdot 2000 \cdot 0^2 - \tfrac{1}{2} \cdot 2000 \cdot 30^2$$

$$-150 f = -900000$$

$$f = 6000 \, N$$

According to the work-energy theorem, the amount of work done on the car is equal to the change in its mechanical energy which in this case is its change in kinetic energy. Since the only force acting on the car is friction, all the work can be attributed to the frictional force.

It is important to realize, for this problem, that the force of friction is against the direction of motion, and thus cosθ = -1.

Problem:
A 20kg child lifts his .5 kg ball off the floor to put it away on his bookshelf 1.5 meters above the ground. How much work has he done?

Solution: The work done is equal to the change in potential energy of the book.

$$W = \Delta PE$$
$$W = mgh_{final} - mgh_{initial}$$
$$W = mg(h_{final} - h_{initial}) = .5 \cdot 9.8 \cdot 1.5 = 7.35J$$

Skill 9.4 Determining power and efficiency as they relate to work and energy in a variety of situations

Skill 9.1 discusses power, work and energy as they relate to mechanical systems. The concept of power as the rate at which work is done or energy is expended is valid for other kinds of systems as well. For instance, the power expended by a light bulb is typically given in Watts. For a discussion of power in heat radiation and in electrical circuits see **Skills 11.7, 13.7** and **15.5**.

Efficiency (ε) essentially measures the output power compared to the input power. The efficiency of a system is a dimensionless number which is usually expressed as a percentage.

Example:
What is the efficiency of a 100 Watt bulb that puts out 75 Watts in the form of light? How much is the remaining energy and what form does it take?
The efficiency of the bulb is 75/100 or 75%. The remaining energy, 25 Watts, goes into thermal energy or the heat the bulb creates.

Another way to define efficiency is to say that it is the percentage of energy that is used to do work on a system.

$$\text{Efficiency:} \quad \varepsilon = \frac{\text{work}(W)}{\text{energy}(E)}$$

Example:
An engine generates 300 J of energy. However, only 80 J of that energy is used to do work. What is the efficiency of the engine?

$$\varepsilon = \frac{\text{work}(W)}{\text{energy}(E)} \Rightarrow \varepsilon = \frac{80J}{300J} \Rightarrow 26.67\%$$

For a discussion of the efficiency of heat engines see **Skill 11.3**.

Skill 9.5 Analyzing energy conversions in everyday-life situations

TEACHER CERTIFICATION STUDY GUIDE

Many forms of energy exist all around us. Energy is defined as the ability to do work. If you are able to measure how much work an object does, or how much heat is exchanged, you can determine the amount of energy that is in a system. Energy and work are measured in Joules. As the law of conservation of energy states, energy can be neither created nor destroyed however, it can be transformed from one form to another. Energy cannot be truly "lost" then, although energy may be wasted or not used to perform work.

Some typical forms of energy are mechanical, heat, sound, electrical, light, chemical, nuclear, and magnetic. Energy can be transformed from mechanical to heat by friction. Additionally, mechanical kinetic energy can combine with magnetic energy, to transform into electrical energy, as when a magnet is spun inside a metal coil. Electrical energy is transformed into light and heat energy when light bulb is turned on and the filament begins to glow. A firefly uses phosphorescence to transform chemical energy into light energy.

Within mechanical energy, energy can transform between potential and kinetic repeatedly as is the case with a pendulum. Energy can also be transformed into matter and vice versa. The equation $E=mc^2$ quantifies the relationship between matter and energy. The conversion of mass to other forms of energy can liberate vast amounts of energy, as shown by nuclear reactors and weapons. However, the mass equivalent of a unit of energy is very small, which is why energy loss is not typically measured by weight.

Energy transformations are classified as thermodynamically reversible or irreversible. A reversible process is one in which no energy is dissipated into empty quantum states, or states of energy with increased disorder. The easiest way to explain the concept is to consider a roller coaster car on a track. A reversible energy transformation occurs as the car travels up and down converting potential energy into kinetic and back. Without friction, the transformation is 100% efficient and no energy is wasted, and the transformation is reversible. However, we know that friction generates heat, and that the heat generated cannot be completely recovered as usable energy, which results in the transformation being irreversible.

Skill 9.6 Analyzing systems in which total energy is conserved but mechanical energy is not conserved

The mechanical energy of a system (U) is the sum of its kinetic energy (KE) and its potential energy (PE). The kinetic energy of a particular object in a system is defined as $KE = \frac{1}{2}mv^2$, where m is the mass and v is the velocity of the object. The potential energy is defined as $PE = mgh$, where g is the acceleration due to gravity and h the height from an arbitrarily defined point of zero potential.

A number of phenomena in everyday life demonstrate that, while mechanical energy is conserved to a certain extent, it is never entirely conserved. In billiards, for example, the motion of one ball can be transferred in part or in whole to another ball. If the first ball, traveling at a certain speed, comes to a complete stop after a collision with another ball, that second ball moves away with the same speed as the initial speed of the first. That is to say, the kinetic energy of the first ball was transferred entirely to the second ball. However, this energy visibly decreases as the ball slows down. This is the result of friction.

A pendulum provides a similar example. The mechanical energy of the pendulum varies periodically from purely kinetic energy at the lowest point of the path to purely potential energy at the highest point of the path. Ideally, the pendulum would swing indefinitely with mechanical energy conserved. Nevertheless, empirical observation shows that pendulums lose energy of motion due to dissipative forces and, eventually, come to a complete stop.

Although mechanical energy is not truly conserved in these cases, a more generalized energy conservation principle can be found in the first law of thermodynamics. Here, the change in the total energy of the system is defined as the sum of the change in heat energy and the work done by the system. That is to say, the total energy of a system is conserved unless one of the following takes place: heat is added or taken away from the system, or mechanical energy is added to or taken away from the system.

$$dU_{total} = \delta Q - \delta W$$

δQ and δW, are infinitesimal energy amounts rather than true differentials. Also, since δW is defined as the work done by the system, rather than the work done on the system, a minus sign is used. Based on this equation, it can be seen that if the system is isolated thermally and mechanically (i.e., both δQ and δW are equal to zero), then there is no change in the total internal energy of the system, dU_{total}. Internally, then, mechanical energy and energy due to heating are together conserved: energy may be transferred between potential and kinetic energy, or it may be transferred between mechanical and heat energy.

It is also noteworthy in the context of atomic or subatomic particles that mechanical energy may be lost in the form of light. A group of particles with a given mechanical energy may, upon interaction, lose some of that mechanical energy in the form of emitted photons. A more complete understanding of these processes, however, requires a quantum mechanical approach rather than a classical approach.

COMPETENCY 10.0 UNDERSTAND THE CHARACTERISTICS OF CIRCULAR MOTION, SIMPLE HARMONIC MOTION, AND OTHER PERIODIC MOTION, AND SOLVE PROBLEMS INVOLVING THESE TYPES OF MOTION

Skill 10.1 Applying vector analysis to describe uniform circular motion in radians

Speed remains constant in uniform circular motion. However, since the object is following a curve, the direction of the velocity is changing. Therefore, the object is accelerating. This acceleration is called **centripetal acceleration**.

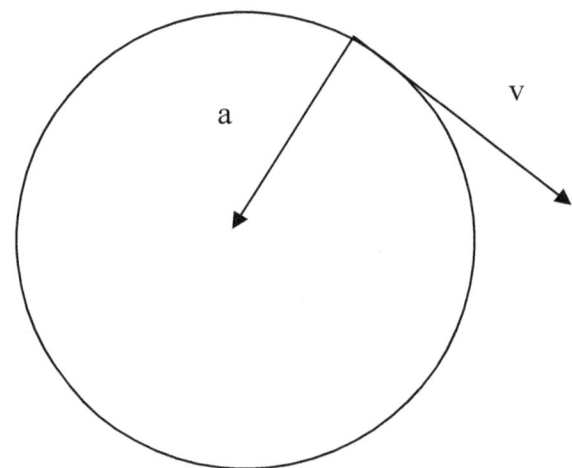

As seen in the above figure, the velocity of the object traveling in a circle is a tangent to the circle. The acceleration of the object is directed towards the center of the circle. Therefore, the object follows a curve.

In order to analyze a circle, it is helpful to use polar coordinates as reference.

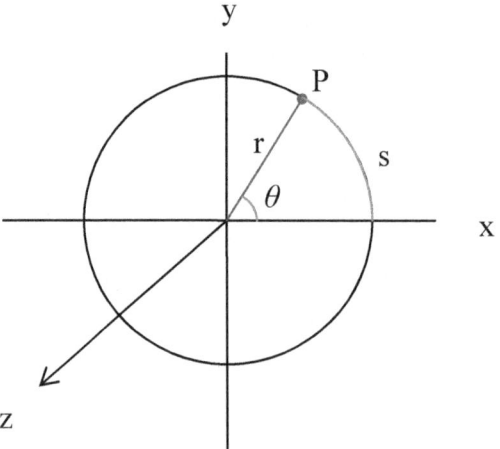

Using polar coordinates (r, θ), Point P on the circle can be defined by the distance from the origin r and the angle with respect to the x-axis θ.

The arc length s can be found using the following equation:

$$s = r\theta$$

where r is the radius and θ is the angular position.

The angular position θ is measure in radians. Note that if the angular position is given in degrees, they must be converted into radians first.

$$1 \text{ revolution} = 360° = 2\pi \text{ radians}$$

Angular velocity
Angular velocity is the angular displacement in a given time period. The instantaneous angular velocity can be found using the equation

$$\omega = \frac{d\theta}{dt}$$

where ω is the angular velocity, θ is the angular position and t is the time.

The relationship between linear velocity (v) and angular velocity (ω)
Imagine there are two revolving bodies that complete one revolution in the same amount of time. However the radius of the orbit of one of the bodies is twice that of the other. It is obvious that the object that has the larger orbit has a greater velocity as it needs to cover a larger distance in the same amount of time. Conversely, the angular velocity of the two bodies is the same as they sweep out equal angles in equal time. Therefore, linear velocity (v) is a function of the radius of a circle (r) for a constant angular velocity (ω).

If $s = r\theta$ then differentiating for time $\frac{ds}{dt} = \frac{dr}{dt}\theta + \frac{d\theta}{dt}r$

This equation can be further simplified as the radius of the circle remains constant, therefore $\frac{ds}{dt} = \frac{d\theta}{dt}r$

Since, $\omega = \frac{d\theta}{dt}$ and $\frac{ds}{dt} = v$, we have $v = \omega r$

Skill 10.2 Determining the magnitude and direction of the force acting on a particle in uniform circular motion

As explained in the previous section, in uniform circular motion the acceleration (a) is directed toward the center of the circular path and is always perpendicular to the velocity, as shown below:

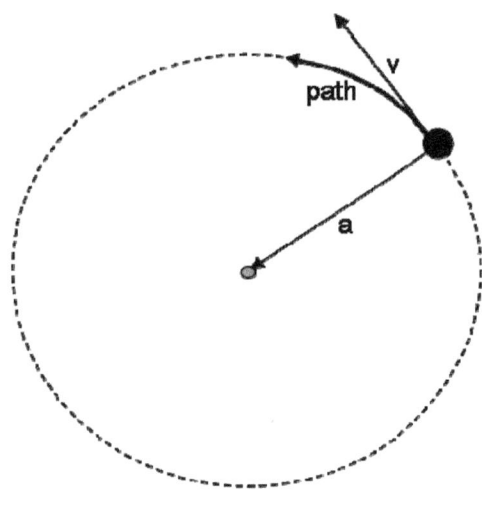

This centripetal acceleration is mathematically expressed as:

$$a = \frac{v^2}{r} = \frac{4\pi^2 r}{t^2}$$

where r is the radius of the circular path and t is the period of the motion or the time taken for the mass to travel once around the circle. The force (F) experienced by the mass (m) is known as centripetal force and is always directed towards the center of the circular path. It has constant magnitude given by the following equation:

$$F = ma = m\frac{v^2}{r}$$

Problem:
A car of mass 2000 Kg travels round a curve of radius 100m at a speed of 90 Km/h. What is the magnitude of the centripetal force acting on it?

Solution:
$$90 Km/h = \frac{90 \times 10^3}{60 \times 60} m/s = 25 m/s$$

$$a = \frac{v^2}{r} = \frac{(25)^2}{100} = 6.25 m/s^2$$

$$F = ma = 2000 \times 6.25 \, N = 12500N$$

Skill 10.3 Applying characteristics of circular motion and gravitational force to analyze planetary motion

Planetary motion is a macroscopic phenomenon that involves non-relativistic speeds (i.e., speeds much less than the speed of light). As a result, classical (Newtonian) mechanics are usually sufficient for analyzing and predicting planetary motion. The most basic problem involves revolution of a planet around a much larger body such as a star. Revolution of a satellite around a planet involves the exact same approach in terms of the applied concepts and solution procedure. The motion of planets, such as the Earth, often involves orbits that cannot be precisely modeled using simple circular motion. Earth's orbit, for example, is more elliptical than circular. Nevertheless, circular motion can be a reasonable first approximation. Gravity is the fundamental force of greatest import on the planetary scale.

Gravitational force
The force between any two point masses, m_1 and m_2, is equal to the product of the two masses and the gravitational constant, G, divided by the square of the distance r between the masses.

$$F_G = G\frac{m_1 m_2}{r^2}$$

Spherical (or nearly spherical) objects such as stars and planets act (approximately) like point masses and may be treated using the above force equation.

Circular motion
For an object of mass m to undergo uniform circular motion with a radius r and velocity v, a centripetal force is required. The expression for this force is given in the equation below.

$$F_C = \frac{mv^2}{r}$$

Planetary motion
The motion of planets, to a good first approximation, can be analyzed in terms of the above-mentioned concepts of gravitational force and circular motion. Choosing one mass (usually the larger mass) as the center of a frame of reference, the motion of the second mass may be modeled as circular with gravitation serving as the centripetal force. Given the masses of the two bodies, the relationship of the velocity to the radius of motion can be established.

$$F_C = F_G = \frac{m_2 v^2}{r} = G\frac{m_1 m_2}{r^2}$$
$$v^2 r = Gm_1$$

In the above expressions, m_1 is chosen as the larger, central body and m_2 as the orbiting body. If the velocity is expressed as the circumference of the circle, $2\pi r$, divided by the period of orbit, T, then the time required for the orbiting body to revolve once around the central body may be calculated.

$$T^2 = \frac{4\pi^2}{Gm_1} r^3$$

Problem: Calculate the period of revolution of Mercury, assuming its radius of circular motion is 57.9×10^9 m.

Solution: Use estimated astronomical data and the gravitational constant along with the above equation.
Mass of the sun: 1.99×10^{30} kg
Gravitational constant: 6.67×10^{-11} m^3 s^{-2} kg^{-1}

$$T = \sqrt{\frac{4\pi^2}{(6.67 \times 10^{-11} \, m^3 s^{-2} kg^{-1})(1.99 \times 10^{30} \, kg)} (57.9 \times 10^9 \, m)^3}$$

$T = 7.60 \times 10^6 \, s = 0.241 y$

Skill 10.4 Analyzing the relationships among displacement, velocity, and acceleration in simple harmonic motion (e.g., simple pendulum, mass on a spring)

Simple harmonic (sinusoidal) motion involves a cyclical exchange of kinetic energy and potential energy as observed in a simple pendulum or a mass on a spring. The relationships among the various parameters of a system displaying simple harmonic motion depends on the type of system being examined. Once the displacement is known, the velocity and acceleration of the object undergoing harmonic motion can be calculated by calculating the first derivative with respect to time (for velocity) or the second derivative with respect to time (for acceleration).

The examples of the mass on a spring and the simple pendulum represent harmonic motion in one dimension and two dimensions, respectively. Each can be treated in a similar manner, although the details vary slightly. It can be shown, however, that the pendulum acts just like the mass on a spring when the displacement angle is small.

Linear oscillator (mass on a spring)
Given the frequency of oscillation f for a mass on a spring (a linear oscillator), along with the concomitant period $T = 1/f$, the displacement of the mass undergoing harmonic (sinusoidal) motion can be written as follows.

$$x(t) = x_{max} \cos(2\pi f t + \phi) = x_{max} \cos(\omega t + \phi)$$

Alternatively, $2\pi f$ can be written as the angular frequency ω. The coefficient x_{max} is the maximum displacement of the mass, and the term Φ is a phase constant that determines the position of the mass at time $t = 0$. If $x(t)$ is differentiated once with respect to time, the velocity of the mass is revealed.

$$v(t) = \frac{\partial x(t)}{\partial t} = -\omega\, x_{max} \sin(\omega t + \phi)$$

Comparing the expressions for displacement and velocity, we can see that the velocity is maximized when the displacement is zero (all kinetic energy), and the displacement is maximized when the velocity is zero (all potential energy); that is, the displacement and velocity are 90^0 out of phase. The acceleration can be calculated by differentiating $v(t)$ with respect to time.

$$a(t) = \frac{\partial v(t)}{\partial t} = -\omega^2\, x_{max} \cos(\omega t + \phi) = -\omega^2 x(t)$$

The acceleration, as shown above, is in phase with the displacement. Applying Newton's second law of motion leads to Hooke's law, which relates the restoring force on the mass to the displacement $x(t)$.

$$F = ma = -m\omega^2 x(t)$$

This equation may be expressed in terms of the so-called spring constant k, which is defined as $m\omega^2$.

$$F = ma = -k\, x(t)$$

Simple pendulum

The simple pendulum model provides a reasonably accurate representation of pendulum motion, especially in the case of small angular amplitude.

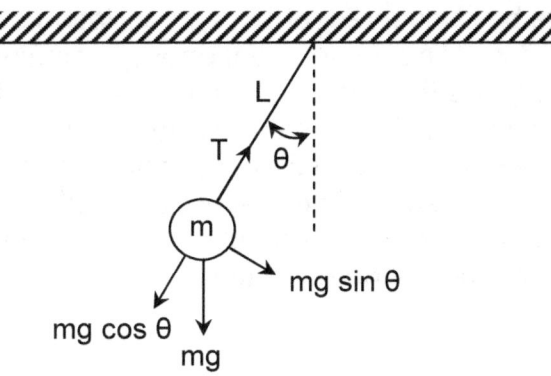

The restoring force F in pendulum motion is expressed as a component of the gravitational force mg perpendicular to the length of the string and is given by

$$F = -mg\sin\theta$$

The negative sign results from the force having a direction opposite to the displacement. The tension T on the string and the portion of the gravitational force in the opposite direction of T cancel one another.

In the case of small θ, sin θ is approximately equal to θ. The arc length traveled by the pendulum, s, is equal to the product of the length L of the string and the angle θ. Thus, the following expression can be derived.

$$F \approx -mg\theta = -mg\frac{s}{L} = -\frac{mg}{L}s$$

In the above equation, the force is shown to be of the same form as the linear harmonic oscillator, having, in this case, a "spring constant" of mg/L. As a result, the expressions found for the displacement, velocity and acceleration in the case of the linear oscillator can also be used here (in the case of small θ), where the frequency ω is replaced as follows.

$$\omega = \sqrt{\frac{g}{L}}$$

Skill 10.5   Analyzing energy interactions in oscillating systems

Though we know total energy is always conserved, there are many physical everyday examples in which we can see how kinetic energy is lost when it is converted to potential energy and vice versa. Recall that kinetic energy is energy possessed by an object due to motion and potential energy is that stored by an object due to position associated with a force.

Total energy must always be conserved, but in an oscillating system that energy is alternately in the form of kinetic and potential energy. Imagine, for instance, a mass connected to a spring. As the spring is compressed, elastic potential energy is stored within the spring. Specifically, when the spring is maximally compressed it also has maximum potential energy and minimum kinetic energy (zero kinetic energy). When the spring is released, this potential energy converts back into kinetic energy as the mass travels outward. This conversion back and forth happens continually as the spring oscillates.

For a mass on a spring with spring constant k, amplitude of oscillation A, angular frequency ω, time t, and phase Φ:

$$\text{Kinetic energy} = \tfrac{1}{2} k A^2 \sin^2(\omega_0 t + \Phi)$$

$$\text{Potential energy} = \tfrac{1}{2} k A^2 \cos^2(\omega_0 t + \Phi)$$

$$\text{Total energy} = \tfrac{1}{2} k A^2$$

We know that sine and cosine are 90° out of phase. Thus, this is a mathematical statement that when potential energy is maximal, kinetic energy is minimum. Further, kinetic energy plus potential energy is always equal to a constant, that is, total energy in the system is always conserved.

Skill 10.6 Solving equilibrium problems involving torques

Torque is rotational motion about an axis. It is defined as $\tau = L \times F$, where L is the lever arm. The length of the lever arm is calculated by measuring the perpendicular line drawn from the line of force to the axis of rotation.

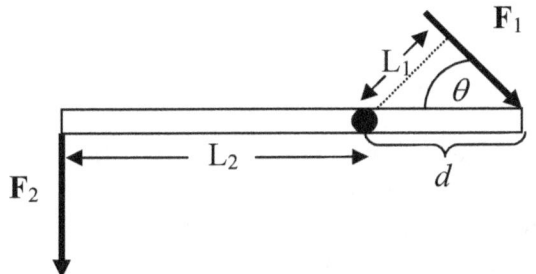

By convention, torques that act in a clockwise direction are considered negative and those in a counterclockwise direction are considered positive. In order for an object to be in equilibrium, the sum of the torques acting on it must be zero.

The equation that would put the above figure in equilibrium is $F_1 L_1 = -F_2 L_2$ (please note that in this case $L_1 = d\sin\theta$).

Examples:
Some children are playing with the spinner below, when one young boy decides to pull on the spinner arrow in the direction indicated by F_1. How much torque does he apply to the spinner arrow?

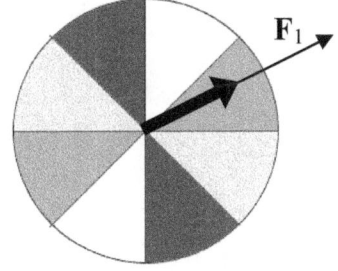

$\tau = L \times F$, but L=0 directly through the perpendicular the pivot point is 0). torque to the

because the line of force goes axis of rotation (i.e. the distance from the line of force to Therefore the boy applies no spinner.

Problem: In the system diagrammed below, find out what the magnitude of F_1 must be in order to keep the system in equilibrium.

Solution: For an object to be in equilibrium the forces acting on it must be balanced. This applies to linear as well as rotational forces known as moments or torques.

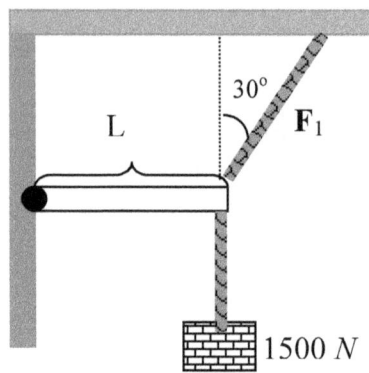

$$F_1 L = -F_2 L$$
$$F_{1x} L + F_{1y} L = -(-1500L) \ldots \text{but } F_{1x} = 0 \text{*}$$
$$F_{1y} L = 1500 L$$
$$F_1 \cos 30 L = 1500 L$$
$$(0.866) F_1 = 1500$$
$$F_1 = 1732 N$$

the effect of $F_{1x}=0$ because that portion of the force goes right through the pivot and causes no torque.

Things to remember:
1. When writing the equation for a body at equilibrium, the point chosen for the axis of rotation is arbitrary.
2. The center of gravity is the point in an object where its weight can be considered to act for the purpose of calculating torque.
3. The lever arm, or moment arm, of a force is calculated as the **perpendicular** distance from the line of force to the pivot point/axis of rotation.
4. Counterclockwise torques are considered positive while clockwise torques are considered negative.

Skill 10.7 Applying the relationship between torque and angular acceleration to solve problems

The relationship between angular acceleration α, moment of inertia I, and torque τ in any situation is given by the rotational analog of Newton's second law of motion:

$$\tau = I\alpha$$

Problem:
The angular velocity of a disc changes steadily from 0 to 6 rad/s in 200 ms. If the moment of inertia of the disc is 10 Kgm^2 what is the torque applied to the disc?

Solution:

The constant angular acceleration of the disc = $\dfrac{6}{200 \times 10^{-3}} = 30 rad/s^2$

Thus torque applied = $10 \times 30 = 300 N.m$.

Problem: A disk of mass 2 Kg and Radius 0.5m is mounted on a horizontal frictionless axle. A block of mass 1.5 Kg is hung from a massless string wrapped around the disk that does not slip. Find the acceleration of the block, the tension in the string and the angular acceleration of the disk.

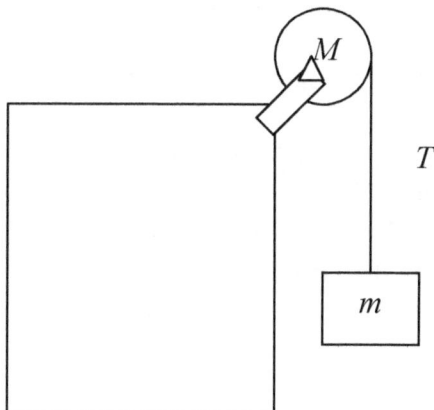

Solution: The forces on the block are the tension T in the string and the weight of the block. The acceleration a of the block is given by Newton's second law:
$$mg - T = ma$$

The torque on the disk of radius R is TR and the moment of inertia of the disk is $\dfrac{1}{2}MR^2$.

PHYSICS 96

Thus, applying Newton's second law in angular form we get

$$TR = \frac{1}{2}MR^2\left(\frac{a}{R}\right)$$

since the string does not slip and the linear acceleration of the disk is the same as the acceleration of the block. Combining both equations,

$$a = \frac{2T}{M} = \frac{2(mg - ma)}{M}; \Rightarrow a(M + 2m) = 2mg; \Rightarrow a = \frac{2mg}{M + 2m} = \frac{2 \times 1.5}{2 + 1.5 \times 2} \times 9.8 = 5.9 m/s^2$$

Using the second equation, the tension in the string is given by

$$T = \frac{1}{2}Ma = 0.5 \times 2 \times 5.9 = 5.9 N$$

The angular acceleration of the disk is given by

$$\alpha = \frac{a}{R} = \frac{5.9}{0.5} = 11.8 rad/s^2$$

Skill 10.8 Applying the law of conservation of angular momentum to describe phenomena and solve problems

The major concepts of linear motion are duplicated in rotational motion with linear displacement replaced by the angle of displacement.

Angular velocity ω = angular displacement / time

Also, the linear velocity v of a rolling object can be written as $v = r\omega$.

One important difference in the equations relates to the use of mass in rotational systems. In rotational problems, not only is the mass of an object important but also its location. In order to include the spatial distribution of the mass of the object, a term called **moment of inertia** is used, $I = m_1 r_1^2 + m_2 r_2^2 + \cdots + m_n r_n^2$

Example:

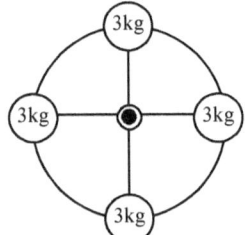

If the radius of the wheel on the left is 0.75m, what is its moment of inertia?

$$I = 3 \cdot 0.75^2 + 3 \cdot 0.75^2 + 3 \cdot 0.75^2 + 3 \cdot 0.75^2 = 6.75$$

Note: $I_{Sphere} = \frac{2}{5}mr^2$, $I_{Hoop/Ring} = mr^2$, $I_{disk} = \frac{1}{2}mr^2$

A related concept is the radius of gyration (k), which is the average distance of mass from the center of an object.

$$k_{Sphere} = \sqrt{\frac{2}{5}}r, \quad k_{Hoop/Ring} = r, \quad k_{disk} = \frac{r}{\sqrt{2}}.$$

As you can see $I = mk^2$

Angular momentum (L), and rotational kinetic energy (KE_r), are therefore defined as follows: $L = I\omega, \quad KE_r = \frac{1}{2}I\omega^2$

Unless a net torque acts on a system, the angular momentum remains constant in both magnitude and direction. This can be used to solve many different types of problems including ones involving satellite motion.

Example:
A planet of mass m is circling a star in an orbit like the one below. If its velocity at point A is 60,000m/s, and $r_B = 8\, r_A$, what is its velocity at point B?

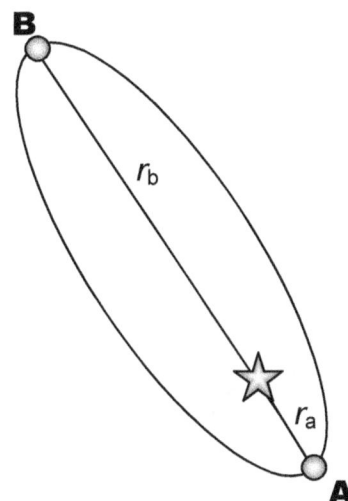

$$I_B\omega_B = I_A\omega_A$$
$$mr_B^2\omega_B = mr_A^2\omega_A$$
$$r_B^2\omega_B = r_A^2\omega_A$$
$$r_B^2 \frac{v_B}{r_B} = r_A^2 \frac{v_A}{r_A}$$
$$r_B v_B = r_A v_A$$
$$8r_A v_B = r_A v_A$$
$$v_B = \frac{v_A}{8} = 7500 m/s$$

Skill 10.9 Applying the law of conservation of energy to systems involving rotational motion

As with all systems, energy is conserved unless the system is acted on by an external force. In situations involving rotational motion, the rotational kinetic energy is also a component of the total energy of the system and is given by $\frac{1}{2}I\omega^2$ where I is the moment of inertia of the system and ω is its angular velocity. (For an extended discussion of these concepts see the previous section **Skill 10.8**).

Problem:
A uniform ball of radius r and mass m starts from rest and rolls down a frictionless incline of height h. When the ball reaches the ground, how fast is it going? The moment of inertia of the ball is given by $\frac{2}{5}mr^2$.

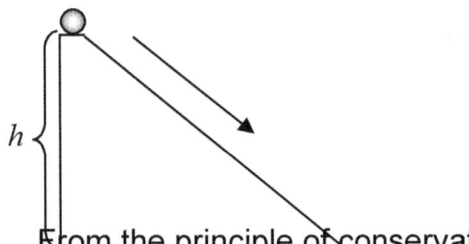

From the principle of conservation of energy we have:

$$PE_{initial} + KE_{rotational/initial} + KE_{linear/initial} = PE_{final} + KE_{rotational/final} + KE_{linear/final}$$

$$mgh + 0 + 0 = 0 + \frac{1}{2}I\omega_{final}^2 + \frac{1}{2}mv_{final}^2 \rightarrow mgh = \frac{1}{2} \cdot \frac{2}{5}mr^2\omega_{final}^2 + \frac{1}{2}mv_{final}^2$$

$$mgh = \frac{1}{5}mr^2(\frac{v_{final}}{r})^2 + \frac{1}{2}mv_{final}^2 \rightarrow mgh = \frac{1}{5}mv_{final}^2 + \frac{1}{2}mv_{final}^2$$

$$gh = \frac{7}{10}v_{final}^2 \rightarrow v_{final} = \sqrt{\frac{10}{7}gh}$$

COMPETENCY 11.0 UNDERSTAND AND APPLY THE PRINCIPLES AND LAWS OF THERMODYNAMICS

Skill 11.1 Analyzing systems in terms of heat energy, internal energy, and work

The **internal energy** of a material is the **sum of the total kinetic energy of its molecules and the potential energy of interactions between those molecules**. Total kinetic energy includes the contributions from translational motion and other components of motion such as rotation. The potential energy includes energy stored in the form of resisting intermolecular attractions between molecules.

The **enthalpy** (H) of a material is the **sum of its internal energy and the mechanical work** it can do by driving a piston. A change in the enthalpy of a substance is the total energy change caused by adding/removing heat at constant pressure.

The first law of thermodynamics

The first law of thermodynamics is a general expression of the conservation of energy wherein heat transfer is considered to be a form of energy transfer in addition to work done on or by a system. The internal energy U of a closed system is related to the heat energy Q in the system and the work W done by the system through the following equation.

$$dU = \delta Q - \delta W$$

Thus, the incremental change in the internal energy of the system is equal to the difference between the incremental amount of heat gained by the system and the incremental amount of work done by the system.

Skill 11.2 Applying the first law of thermodynamics to analyze energy conversions in a variety of situations (e.g., ideal gas, electrical circuit, mechanical systems, optical systems, acoustic systems)

The first law of thermodynamics applies in a number of situations. Ideally, in many simple models of various phenomena, undesirable effects such as friction, resistance and absorption are ignored. In reality, however, these are unavoidable consequences of the nature of the phenomena. In the context of **electrical circuits**, collisions in wires or devices impedes the flow of electrons and can convert mechanical energy, which originates from an applied electric or magnetic field, into heat energy. If the heat is dissipated from the circuit then that heat loss represents energy loss, implying a decrease in the circuit's efficiency since less energy is available to do work.

A similar situation exists in the case of **mechanical systems**. The motion of a moving object can be impeded by the resistance of air, the surfaces of other objects or the viscosity of a liquid. Since energy is conserved, the decrease in mechanical energy in the system is balanced by an increase of heat energy. This heat is energy transferred from motion (kinetic energy) through friction. A swinging pendulum, for example, has alternating kinetic and potential energy, but, over time, this mechanical energy is lost to heat energy by way of friction. If the whole system is thermally and mechanically isolated (i.e., $dU = 0$), eventually all the mechanical energy of the pendulum will become heat energy. As a side note, it is this phenomenon that, in the context of cosmology, is referred to as "heat death." Some physicists believe that all the "useful" energy of the universe will eventually become heat energy, resulting in a "winding down" of the universe.

In the case of **optical systems**, the energy contained in incident light can be lost to heat through absorption. On a microscopic level, as described by quantum mechanics, incident photons can be absorbed by a material, causing an excitation of a particular atom or number of atoms. Although these excited atoms may relax to lower energy states through the emission of photons (thus maintaining the energy in the form of light), the relaxation process may also occur through emission of a phonon, or vibrational mode. These phonons are responsible for heat conduction in a material.

For **acoustics**, sound waves can lose energy to heating as well. As variations in pressure occur, energy in acoustic vibrations can be transferred to heat energy in the material. Again, the total energy of the system must remain constant, and therefore the sum of the mechanical and thermal energies cannot change as long as the system remains closed.

In the case of an **ideal gas**, which obeys the ideal gas law $PV=nRT$, work done on or by the system result in a transfer of energy available to do work into heat energy. A particular case would be a confined ideal gas; if the volume V of the gas is decreased, meaning work is done on the system, either the pressure P of the gas increases (in which case the gas acquires energy available to do mechanical work) or the heat energy of the gas (as measured by the temperature T) increases. Alternatively, some combination of changes in these properties takes place. If the gas is allowed to cool, venting heat energy out of the gas, then the total energy of the gas decreases, meaning that, in this case, less energy is available for the gas to do work.

Skill 11.3 Demonstrating an understanding of the second law of thermodynamics and of how the entropy of a system changes in a variety of situations (e.g., an ice cube melting, a gas cooled at constant volume, Carnot cycle)

To understand the **second law of thermodynamics**, we must first understand the concept of entropy. Entropy is the transformation of energy to a more disordered state and is the measure of how much energy or heat is available for work. The greater the entropy of a system, the less energy is available for work. The simplest statement of the second law of thermodynamics is that the entropy of an isolated system not in equilibrium tends to increase over time. The entropy approaches a maximum value at equilibrium. Below are several common examples in which we see the manifestation of the second law.

- The diffusion of molecules of perfume out of an open bottle
- Even the most carefully designed engine releases some heat and cannot convert all the chemical energy in the fuel into mechanical energy
- A block sliding on a rough surface slows down
- An ice cube sitting on a hot sidewalk melts into a little puddle; we must provide energy to a freezer to facilitate the creation of ice

When discussing the second law, scientists often refer to the "arrow of time". This is to help us conceptualize how the second law forces events to proceed in a certain direction. To understand the direction of the arrow of time, consider some of the examples above; we would never think of them as proceeding in reverse. That is, as time progresses, we would never see a puddle in the hot sun spontaneously freeze into an ice cube or the molecules of perfume dispersed in a room spontaneously re-concentrate themselves in the bottle. The above-mentioned examples are **spontaneous** as well as **irreversible**, both characteristic of increased entropy. Entropy change is zero for a complete cycle in a **reversible process**, a process where infinitesimal quasi-static changes in the absence of dissipative forces can bring a system back to its original state without a net change to the system or its surroundings. All real processes are irreversible. The idea of a reversible process, however, is a useful abstraction that can be a good approximation in some cases.

A quantitative measure of entropy S is given by the statement that the change in entropy of a system that goes from one state to another in an isothermal and reversible process is the amount of heat absorbed in the process divided by the absolute temperature at which the process occurs.

$$\Delta S = \frac{\Delta Q}{T}$$

Stated more generally, the entropy change that occurs in a state change between two equilibrium states A and B via a reversible process is given by

$$\Delta S_{A \to B} = \int_A^B \frac{dQ}{T}$$

<u>Problem:</u> What is the change in entropy of a cube of ice of mass 30g which melts at temperature 0C? The latent heat of fusion of ice is 334 KJ/Kg.

<u>Solution:</u> The amount of heat absorbed by the ice cube =
$30 \times 10^{-3} \times 334 KJ = 10020 J$.

Thus change in entropy = $(10020/273) J/K$ = 36.7 J/K

The second law of thermodynamics may also be stated in the following ways:
1. No machine is 100% efficient.
2. Heat cannot spontaneously pass from a colder to a hotter object.

If we consider a **heat engine** that absorbs heat Q_h from a hot reservoir at temperature T_h and does work W while rejecting heat Q_c to a cold reservoir at a lower temperature T_c, $Q_h - Q_c = W$. The efficiency of the engine is the ratio of the work done to the heat absorbed and is given by

$$\varepsilon = \frac{W}{Q_h} = \frac{Q_h - Q_c}{Q_h} = 1 - \frac{Q_c}{Q_h}$$

It is impossible to build a heat engine with 100% efficiency, i.e. one where $Q_c = 0$.

Carnot described an ideal reversible engine, the **Carnot engine**, that works between two heat reservoirs in a cycle known as the **Carnot cycle** which consists of two isothermal (12 and 34) and two adiabatic processes (23 and 41) as shown in the diagram below.

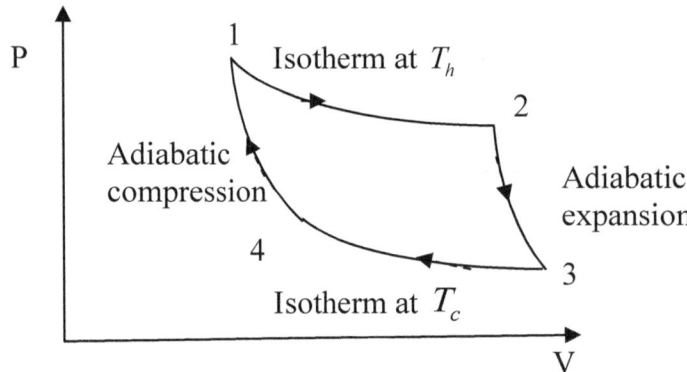

The efficiency of a Carnot engine is given by $\varepsilon = 1 - \dfrac{T_c}{T_h}$ where the temperature values are absolute temperatures. This is the highest efficiency that any engine working between T_c and T_h can reach.

According to **Carnot's theorem**, no engine working between two heat reservoirs can be more efficient than a reversible engine. All such reversible engines have the same efficiency.

Skill 11.4   Analyzing characteristics of temperature and temperature scales

Heat is generally measured in terms of temperature, a measure of the average internal energy of a material. Temperature is an **intensive property**, meaning that it does not depend on the amount of material. Heat content is an **extensive property** because more material at the same temperature will contain more heat. The relationship between the change in heat energy of a material and the change in its temperature is given by $\Delta Q = mC\Delta T$, where ΔQ is the change in heat energy, m is the mass of the material, ΔT is the change in temperature and C is the specific heat which is characteristic of a particular material.

The **zeroth law of thermodynamics** generally deals with bodies in thermal equilibrium with each other and is the basis for the idea of temperature. Most commonly, the law is stated as, "If two thermodynamic systems are in thermal equilibrium with a third, they are also in thermal equilibrium with each other." Said another way, this very basic law simply states that if object A is same temperature as object B, and object C is the same temperature as object B, then object A and C are also the same temperature.

The **third law of thermodynamics** defines absolute zero of temperature as the point at which a system is perfectly ordered and contains no thermal energy. One common way of stating it is, "The entropy of a perfect crystal of an element at the absolute zero of temperature is zero." The law states that, as a system approaches absolute zero, it will become more orderly. Derivative from this law is the fact that the entropy of a given system near absolute zero is dependent only upon the temperature. At a temperature of absolute zero, there is no thermal energy or heat, so there is no molecular motion.

There are four generally recognized temperature scales, Celsius, Fahrenheit, Kelvin and Rankine. The Kelvin and Rankine scales are absolute temperature scales corresponding to the Celsius and Fahrenheit scales, respectively. Absolute temperature scales have a zero reading when the temperature reaches absolute zero (the theoretical point at which no thermal energy exists). The absolute temperature scales are used often in physics and chemistry.

To convert between Celsius and Fahrenheit, use the following relationship:

$$x \,°F = (5/9)(x - 32) \,°C$$

To convert to the absolute temperature scales, use the appropriate conversion below:

$$x \,°F = x + 459.67 \,°R$$

$$x \,°C = x + 273.15 \,K$$

Note that the size of each degree on the Fahrenheit/ Rankine scale is smaller than the size of a degree on the Celsius/Kelvin scale.

Skill 11.5 Solving problems involving thermal expansion and thermal contraction of solids, liquids, and ideal gases

Solids and liquids
Most materials expand when heated with a change in dimension proportional to the change in temperature. If we consider a long rod of length L that increases in length by ΔL when heated, the fractional change in length $\Delta L / L$ is directly proportional to the change in temperature ΔT.

$$\Delta L / L = \alpha . \Delta T$$

The constant of proportionality α is known as the **coefficient of linear expansion** and is a property of the material of which the rod is made.

Problem: The temperature of an iron rod 10 meters long changes from $-3°C$ to $12°C$. If iron has a coefficient of linear expansion of 0.000011 per $°C$, by how much does the rod expand?

Solution: The length of the rod L = 10 meters.
Change in temperature $\Delta T = 12°C - (-3°C) = 15°C$
Change in length of the rod $\Delta L = 0.000011 \times 10 \times 15 = .00165$ meters

If instead of a rod, we consider an area A that increases by ΔA when heated, we find that the fractional change in area is proportional to the change in temperature ΔT. The proportionality constant in this case is known as the **coefficient of area expansion** and is related to the coefficient of linear expansion as demonstrated below.

If A is a rectangle with dimensions L_1 and L_2, then

$$A + \Delta A = (L_1 + \Delta L_1)(L_2 + \Delta L_2)$$
$$= (L_1 + \alpha L_1 \Delta T)(L_2 + \alpha L_2 \Delta T)$$
$$= L_1 L_2 + 2\alpha L_1 L_2 \Delta T + \alpha^2 (\Delta T)^2$$

Ignoring the higher order term for small changes in temperature, we find that

$$\Delta A = 2\alpha A \Delta T = \gamma A \Delta T$$

Thus the coefficient of area expansion $\gamma = 2\alpha$.

Following the same procedure as above, we can show that the change in volume of a material when heated may be expressed as

$$\Delta V = 3\alpha V \Delta T = \beta V \Delta T$$

where the **coefficient of volume expansion** $\beta = 3\alpha$.

Problem: An aluminum sphere of radius 10cm is heated from $0°C$ to $25°C$. What is the change in its volume? The coefficient of linear expansion of aluminum is 0.000024 per $°C$.

Solution: Volume V of the sphere = $\frac{4}{3}\Pi r^3 = \frac{4}{3} \times 3.14 \times 1000 cm^3 = 4186.67 cm^3$

Change in volume of the sphere = $3 \times 0.000024 \times 4186.67 \times 25 = 7.54 cm^3$

For a liquid, the only meaningful expansion parameter is the coefficient of volume expansion. Most liquids expand when heated, a notable exception being water between $0°C$ and $4°C$. This is the reason bodies of water freeze from the top with and allow aquatic life to flourish even in cold temperatures.

Ideal gases

The pressure, temperature and volume relationships for an ideal gas (a gas described by the assumptions of the kinetic molecular theory listed above) are given by the following gas laws:

Boyle's law states that the volume of a fixed amount of gas at constant temperature is inversely proportional to the gas pressure, or:

$$V \propto \frac{1}{P}. \text{ or } PV = \text{constant}$$

Problem: A 15 liter container contains gas at a pressure of 5 atmospheres. If the volume of the container is reduced to 3 liters and the temperature remains unchanged, what will be the pressure in the container?

Solution: Since PV is constant by Boyle's law, the product of the initial pressure and volume is equal to the product of the final pressure and volume, i.e.
$$P_1V_1 = P_2V_2$$

Therefore, $P_2 = P_1V_1/V_2 = 15 \times 5/3 \, atm. = 25 atm.$

Gay-Lussac's law states that the pressure of a fixed amount of gas in a fixed volume is proportional to absolute temperature, or:
$$P \propto T.$$

Charles' law states that the volume of a fixed amount of gas at constant pressure is directly proportional to absolute temperature, or:
$$V \propto T.$$

Problem: A container has 25 liters of gas at $0°C$. What will the volume of the gas be if the temperature is raised to $100°C$ at constant pressure?

Solution: From Charles' law we know that $V_1/T_1 = V_2/T_2$.

Thus $V_2 = V_1T_2/T_1 = 25 \times 373/273 = 34.2 \, liters$.

The **combined gas law** uses the above laws to determine a proportionality expression that is used for a constant quantity of gas:
$$V \propto \frac{T}{P}.$$
The combined gas law is often expressed as an equality between identical amounts of an ideal gas at two different states ($n_1=n_2$):
$$\frac{P_1V_1}{T_1} = \frac{P_2V_2}{T_2}.$$

Avogadro's hypothesis states that equal volumes of different gases at the same temperature and pressure contain equal numbers of molecules. **Avogadro's law** states that the volume of a gas at constant temperature and pressure is directly proportional to the quantity of gas, or $V \propto n$ where n is the number of moles of gas.

Avogadro's law and the combined gas law yield $V \propto \dfrac{nT}{P}$. The proportionality constant R--the **ideal gas constant**--is used to express this proportionality as the **ideal gas law**:

$$PV = nRT.$$

The ideal gas law is useful because it contains all the information of Charles's, Avogadro's, Boyle's, and the combined gas laws in a single expression.

Skill 11.6   Analyzing methods of heat transfer (i.e., conduction, convection, and radiation) in practical situations

All heat transfer is the movement of thermal energy from hot to cold matter. This movement down a thermal gradient is a consequence of the second law of thermodynamics. The three methods of heat transfer are listed and explained below.

Conduction: Electron diffusion or photo vibration is responsible for this mode of heat transfer. The bodies of matter themselves do not move; the heat is transferred because adjacent atoms that vibrate against each other or as electrons flow between atoms. This type of heat transfer is most common when two solids come in direct contact with each other. This is because molecules in a solid are in close contact with one another and so the electrons can flow freely. It stands to reason, then, that metals are good conductors of thermal energy. This is because their metallic bonds allow the freest movement of electrons. Similarly, conduction is better in denser solids. Examples of conduction can be seen in the use of copper to quickly convey heat in cooking pots, the flow of heat from a hot water bottle to a person's body, or the cooling of a warm drink with ice.

Convection: Convection involves some conduction but is distinct in that it involves the movement of warm particles to cooler areas. Convection may be either natural or forced, depending on how the current of warm particles develops. Natural convection occurs when molecules near a heat source absorb thermal energy (typically via conduction), become less dense, and rise. Cooler molecules then take their place and a natural current is formed. Forced convection, as the name suggests, occurs when liquids or gases are moved by pumps, fans, or other means to be brought into contact with warmer or cooler masses. Because the free motion of particles with different thermal energy is key to this mode of heat transfer, convection is most common in liquid and gases. Convection can, however, transfer heat between a liquid or gas and a solid. Forced convection is used in "forced air" home heating systems and is common in industrial manufacturing processes. Additionally, natural convection is responsible for ocean currents and many atmospheric events. Finally, natural convection often arises in association with conduction, for instance in the air near a radiator or the water in a pot on the stove.

Radiation: This method of heat transfer occurs via electromagnetic radiation. All matter warmer than absolute zero (that is, all known matter) radiates heat. This radiation occurs regardless of the presence of any medium. Thus, it occurs even in a vacuum. Since light and radiant heat are both part of the EM spectrum, we can easily visualize how heat is transferred via radiation. For instance, just like light, radiant heat is reflected by shiny materials and absorbed by dark materials. Common examples of radiant heat include the way sunlight travels from the sun to warm the earth, the use of radiators in homes, and the warmth of incandescent light bulbs.

Skill 11.7 **Solving problems involving heat capacity, specific heat, heat transfer, heat of fusion, heat of vaporization, and phase changes**

Heat capacity and phase change
A substance's molar **heat capacity** is the heat required to **change the temperature of one mole of the substance by one degree**. Heat capacity has units of joules per mol- kelvin or joules per mol- °C. The two units are interchangeable because we are only concerned with differences between one temperature and another. A Kelvin degree and a Celsius degree are the same size.

The **specific heat** of a substance (also called specific heat capacity) is the heat required to **change the temperature of one gram or kilogram by one degree**. Specific heat has units of joules per gram-°C or joules per kilogram-°C.

The temperature of a material rises when heat is transferred to it and falls when heat is removed from it. When the material is undergoing a phase change (e.g. from solid to liquid), however, it absorbs or releases heat without a corresponding change in temperature. The heat that is absorbed or released during phase change is known as latent heat.

A **temperature vs. heat graph** can demonstrate these relationships visually. One can also calculate the specific heat or latent heat of phase change for the material by studying the details of the graph.

Example: The plot below shows heat applied to 1g of ice at -40C. The horizontal parts of the graph show the phase changes where the material absorbs heat but stays at the same temperature. The graph shows that ice melts into water at 0C and the water undergoes a further phase change into steam at 100C.

The specific heat of ice, water and steam and the latent heat of fusion and vaporization may be calculated from each of the five segments of the graph.

For instance, we see from the flat segment at temperature 0C that the ice absorbs 80 cal of heat. The latent heat L of a material is defined by the equation $\Delta Q = mL$ where ΔQ is the quantity of heat transferred and m is the mass of the material. Since the mass of the material in this example is 1g, the latent heat of fusion of ice is given by $L = \Delta Q / m = 80$ cal/g.

The next segment shows a rise in the temperature of water and may be used to calculate the specific heat C of water defined by $\Delta Q = mC\Delta T$, where ΔQ is the quantity of heat absorbed, m is the mass of the material and ΔT is the change in temperature. According to the graph, ΔQ = 200-100 =100 cal and ΔT = 100-0=100C. Thus, C = 100/100 = 1 cal/gC.

Problem: The plot below shows the change in temperature when heat is transferred to 0.5g of a material. Find the initial specific heat of the material and the latent heat of phase change.

Solution:
Looking at the first segment of the graph, we see that $\Delta Q = 40$ cal and $\Delta T = 120$ C. Since the mass $m = 0.5$g, the specific heat of the material is given by
$C = \Delta Q / (m\Delta T) = 40/(0.5 \times 120) = 0.67$ cal/gC.

The flat segment of the graph represents the phase change. Here
$\Delta Q = 100 - 40 = 60$ cal. Thus, the latent heat of phase change is given by
$L = \Delta Q / m = 60/(0.5) = 120$ cal/g.

Heat transfer
The amount of heat transferred by conduction through a material depends on several factors. It is directly proportional to the temperature difference ΔT between the surface from which the heat is flowing and the surface to which it is transferred. Heat flow H increases with the area A through which the flow occurs and also with the time duration t. The thickness of the material reduces the flow of heat. The relationship between all these variables is expressed as

$$H = \frac{k.t.A.\Delta T}{d}$$

where the proportionality constant k is known as the **thermal conductivity**, a property of the material. Thermal conductivity of a good conductor is close to 1 (0.97 cal/cm.s.^0C for silver) while good insulators have thermal conductivity that is nearly zero (0.0005 cal/cm.s.^0C for wood).

Problem: A glass window pane is 50 cm long and 30 cm wide. The glass is 1 cm thick. If the temperature indoors is $15\,^\circ C$ higher than it is outside, how much heat will be lost through the window in 30 minutes? The thermal conductivity of glass is 0.0025 cal/cm.s.$^\circ C$.

Solution: The window has area A = 1500 sq. cm and thickness d = 1 cm. Duration of heat flow is 1800 s and the temperature difference $\Delta T = 15\,^\circ C$. Therefore heat loss through the window is given by

$$H = (0.0025 \times 1800 \times 1500 \times 15)/1 = 101250 \text{ calories}$$

The amount of energy radiated by a body at temperature T and having a surface area A is given by the Stefan-Boltzmann law expressed as

$$I = e\sigma A T^4$$

where I is the radiated power in watts, e (a number between 0 and 1) is the **emissivity** of the body and σ is a universal constant known as **Stefan's constant** that has a value of $5.6703 \times 10^{-8} W/m^2.K^4$. Black objects absorb and radiate energy very well and have emissivity close to 1. Shiny objects that reflect energy are not good absorbers or radiators and have emissivity close to zero.

A body not only radiates thermal energy but also absorbs energy from its surroundings. The net power radiation from a body at temperature T in an environment at temperature T_0 is given by

$$I = e\sigma A(T^4 - T_0^4)$$

Problem: Calculate the net power radiated by a body of surface area 2 sq. m, temperature $30\,^\circ C$ and emissivity 0.5 placed in a room at a temperature of $15\,^\circ C$.

Solution: $I = 0.5 \times 5.67 \times 10^{-8} \times 2(303^4 - 288^4) = 88$ W

TEACHER CERTIFICATION STUDY GUIDE

Skill 11.8 Describing thermal properties of matter (e.g., solids, liquids, gases) in terms of molecular theory

The relationship between **kinetic energy** and **intermolecular forces** determines whether a collection of molecules will be a gas, liquid, or solid. When a material is heated and experiences a phase change from a solid to liquid or liquid to gas, thermal energy is used to break the intermolecular bonds holding the material together. Similarly, bonds are formed with the release of thermal energy when a material changes its phase during cooling.

In a gas, the energy of intermolecular forces is much weaker than the kinetic energy of the molecules. Kinetic molecular theory is usually applied to gases and is best applied by imagining ourselves shrinking down to become a molecule and picturing what happens when we bump into other molecules and into container walls.

Gas **pressure** results from molecular collisions with container walls. The **number of molecules** striking an **area** on the walls and the **average kinetic energy** per molecule are the only factors that contribute to pressure. A higher **temperature** increases speed and kinetic energy. There are more collisions at higher temperatures, but the average distance between molecules does not change, and thus density does not change in a sealed container.

Kinetic molecular theory explains why the pressure and temperature of gases behave the way they do by making a few assumptions, namely:

1) The energies of intermolecular attractive and repulsive forces may be neglected.
2) The average kinetic energy of the molecules is proportional to absolute temperature.
3) Energy can be transferred between molecules during collisions and the collisions are elastic, so the average kinetic energy of the molecules doesn't change due to collisions.
4) The volume of all molecules in a gas is negligible compared to the total volume of the container.

Strictly speaking, molecules also contain some kinetic energy by rotating or experiencing other motions. The motion of a molecule from one place to another is called **translation**. Translational kinetic energy is the form that is transferred by collisions, and kinetic molecular theory ignores other forms of kinetic energy because they are not proportional to temperature.

PHYSICS

The following table summarizes the application of kinetic molecular theory to an increase in container volume, number of molecules, and temperature:

Effect of an **increase** in one variable with other two constant	Impact on gas: − = decrease, 0 = no change, + = increase						
	Average distance between molecules	Density in a sealed container	Average speed of molecules	Average translational kinetic energy of molecules	Collisions with container walls per second	Collisions per unit area of wall per second	Pressure (P)
Volume of container (V)	+	−	0	0	−	−	−
Number of molecules	−	+	0	0	+	+	+
Temperature (T)	0	0	+	+	+	+	+

Additional details on the kinetic molecular theory may be found at http://hyperphysics.phy-astr.gsu.edu/hbase/kinetic/ktcon.html. An animation of gas particles colliding is located at http://comp.uark.edu/~jgeabana/mol_dyn/

TEACHER CERTIFICATION STUDY GUIDE

SUBAREA III   **ELECTRICITY AND MAGNETISM**

COMPETENCY 12.0   **UNDERSTAND CHARACTERISTICS AND UNITS OF ELECTRIC CHARGE, ELECTRIC FIELDS, ELECTRIC POTENTIAL, AND CAPACITANCE; AND APPLY PRINCIPLES OF STATIC ELECTRICITY TO PROBLEMS INVOLVING COULOMB'S LAW AND ELECTRIC FIELD INTENSITY**

Skill 12.1   **Analyzing the behavior of an electroscope in given situations (e.g., charging by induction and conduction)**

An electroscope is, fundamentally, a charge-sensing device. Although it involves no digital output, it can be a helpful heuristic tool for understanding some basic concepts surrounding positive and negative charges. A simple design for an electroscope involves a glass container with a metal ball outside and a metal contact inside. Connected to the contact is a thin metal leaf that can be deflected by only a small applied force.

The behavior of the metal leaf inside the electroscope provides insight into the presence of charges in the surrounding environment or in the metal portions of the electroscope itself.

If a charged object is brought near (but does not touch) the electroscope (i.e., the metal ball), the free charge carriers in the metal rearrange themselves to minimize the energy of the system. If the object is negatively charged due to an excess of electrons, for example, the otherwise evenly distributed electrons in the metal of the electroscope will reorient such that a net positive charge is maintained near the charged object. This reorientation leads to a net negative charge in the other portion of the electroscope (i.e., inside the glass container), since the electroscope is initially uncharged. The local net charge results in electrical repulsion between the metal contact and the metal leaf. The leaf is then deflected.

If, while the charged object is near, the metal ball is touched by an observer, for example, the electroscope becomes "grounded," and some of the excess charge in certain areas is discharged. This results in the leaf returning to its original undeflected position. When the charged object is removed (after the observer also breaks contact), the result is that the metal leaf remains deflected. This phenomenon results from the excess charge obtained from grounding the electroscope during the time it was in the presence of a charged object. In the case of the above example, an excess of positive charge would remain and would disperse evenly throughout the metal (i.e., evenly very near the surface).

If the electroscope is grounded once more, with the charged object no longer present, the excess charge will be lost and the leaf will return to its undeflected position.

In addition to the above method of charging the electroscope by induction, it may also be charged by rubbing the outside metal with a charged object, such as a glass rod. The conducting metal allows the excess charge on the rod to further distribute itself into a preferred lower-energy situation. This results in a deflection of the leaf that, as with the previous case, can be reversed by grounding the electroscope.

Skill 12.2 Analyzing common electrostatic phenomena (e.g., static cling, lightning, St. Elmo's fire)

Electrostatics involves the study of problems that include only stationary electric charges. Although most matter is electrically neutral on a macroscopic scale, the effects of electrostatic phenomena still can be witnessed in common, everyday experiences.

Perhaps the most common example is so-called **static electricity** which results from a buildup of charge on two separate objects or different regions of the same object. This excess charge, and corresponding electric potential, will repel objects with an excess charge of the same type (i.e., both positive or both negative) and attract objects of dissimilar type (i.e., positive and negative or vice versa). Thus, for example, a glass rod rubbed with a piece of silk acquires a net positive charge due to a loss of some of its negatively charged electrons. A plastic rod rubbed with wool acquires a net negative charge due to a deposit of excess electrons from the wool. These two rods, since they are of opposite charge, will attract one another.

Lightning, while not an electrostatic phenomenon, strictly speaking, results from a similar situation where an imbalance of positive and negative charge is built up in different regions of the earth or sky. This charge imbalance creates an electric potential that can be discharged as lightning.

St. Elmo's fire, named after the patron saint of sailors, is a kind of continuous electric discharge or glow discharge that appears when a gas is exposed to high electrical voltage. It is often seen during thunderstorms when the air molecules are ionized by high voltages and a plasma is created. It is essentially the same phenomenon as the one occurring in fluorescent tubes.

Skill 12.3 Applying Coulomb's law to determine the forces between charges

Any point charge may experience force resulting from attraction to or repulsion from another charged object. The easiest way to begin analyzing this phenomenon and calculating this force is by considering two point charges. Let us say that the charge on the first point is Q_1, the charge on the second point is Q_2, and the distance between them is r. Their interaction is governed by Coulomb's Law which gives the formula for the force F as:

$$F = k \frac{Q_1 Q_2}{r^2}$$

where $k = 9.0 \times 10^9 \frac{N \cdot m^2}{C^2}$ (known as Coulomb's constant)

The charge is a scalar quantity, however, the force has direction. For two point charges, the direction of the force is along a line joining the two charges. Note that the force will be repulsive if the two charges are both positive or both negative and attractive if one charge is positive and the other negative. Thus, a negative force indicates an attractive force.

When more than one point charge is exerting force on a point charge, we simply apply Coulomb's Law multiple times and then combine the forces as we would in any statics problem. Let's examine the process in the following example problem.

Problem: Three point charges are located at the vertices of a right triangle as shown below. Charges, angles, and distances are provided (drawing not to scale). Find the force exerted on the point charge A.

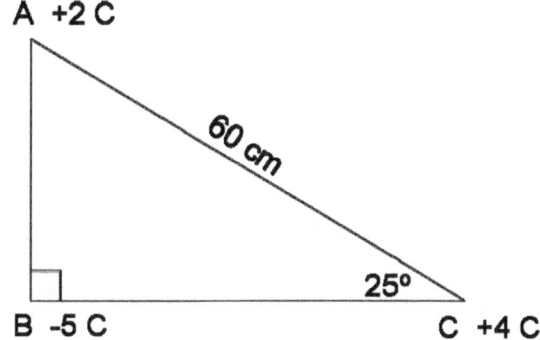

Solution: First we find the individual forces exerted on A by point B and point C. We have the information we need to find the magnitude of the force exerted on A by C.

$$F_{AC} = k\frac{Q_1 Q_2}{r^2} = 9 \times 10^9 \frac{N \cdot m^2}{C^2}\left(\frac{4C \times 2C}{(0.6m)^2}\right) = 2 \times 10^{11} N$$

To determine the magnitude of the force exerted on A by B, we must first determine the distance between them.

$$\sin 25° = \frac{r_{AB}}{60cm}$$
$$r_{AB} = 60cm \times \sin 25° = 25cm$$

Now we can determine the force.

$$F_{AB} = k\frac{Q_1 Q_2}{r^2} = 9 \times 10^9 \frac{N \cdot m^2}{C^2}\left(\frac{-5C \times 2C}{(0.25m)^2}\right) = -1.4 \times 10^{12} N$$

PHYSICS

We can see that there is an attraction in the direction of B (negative force) and repulsion in the direction of C (positive force). To find the net force, we must consider the direction of these forces (along the line connecting any two point charges). We add them together using the law of cosines.

$$F_A^2 = F_{AB}^2 + F_{AC}^2 - 2F_{AB}F_{AC}\cos 75°$$
$$F_A^2 = (-1.4\times 10^{12}\,N)^2 + (2\times 10^{11}\,N)^2 - 2(-1.4\times 10^{12}\,N)(2\times 10^{11}\,N)^2 \cos 75°$$
$$F_A = 1.5\times 10^{12}\,N$$

This gives us the magnitude of the net force, now we will find its direction using the law of sines.

$$\frac{\sin\theta}{F_{AC}} = \frac{\sin 75°}{F_A}$$
$$\sin\theta = F_{AC}\frac{\sin 75°}{F_A} = 2\times 10^{11}\,N\frac{\sin 75°}{1.5\times 10^{12}\,N}$$
$$\theta = 7.3°$$

Thus, the net force on A is 7.3° west of south and has magnitude 1.5 x 10^{12}N. Looking back at our diagram, this makes sense, because A should be attracted to B (pulled straight south) but the repulsion away from C "pushes" this force in a westward direction.

Skill 12.4 Applying principles of electrostatics to determine electric field intensity

Electric fields can be generated by a single point charge or by a collection of charges in close proximity. The electric field generated from a point charge is given by:

$$E = \frac{kQ}{r^2}$$

where E = the electric field

k = $9.0\times 10^9\,\frac{N\cdot m^2}{C^2}$ (Coulomb's constant)

Q = the point charge
r = distance from the charge

Electric fields are visualized with field lines, which demonstrate the strength and direction of an electric field. The electric field around a positive charge points away from the charge and the electric field around a negative charge points toward the charge.

While it's easy enough to calculate and visualize the field generated by a single point charge, we can also determine the nature of an electric field produced by a collection of charge simply by adding the vectors from the individual charges. This is known as the superposition principle. The following equation demonstrates how this principle can be used to determine the field resulting from hundreds or thousands of charges.

$$\vec{E}_{\text{total}} = \sum_i \vec{E}_i = \vec{E}_1 + \vec{E}_2 + \vec{E}_3 \ldots$$

Skill 12.5 Analyzing the vector nature of electric fields and forces

A field is essentially a parameter that has a certain value at each point in space. Fields may have only magnitude (scalar fields) or both magnitude and direction (vector fields). Temperature is an example of a scalar field since it has a certain value that can be measured at any point. However, there is no directionality associated with temperature. A vector field like an electric field, however, is characterized both by a magnitude and a direction. The vector nature of the electric field arises from the fact that it is essentially the force that a unit positive charge would experience if placed in the field. Even though the field exists in the absence of any such actual charge, the magnitude and direction of the field are defined in this way.

Electric field lines, indicating the direction of the field at a point in space, have been discussed in the previous section. The previous section also discusses how multiple electric fields may be combined, like forces, using vector addition. For a review of how to combine vectors graphically or by trigonometric methods, see **Skill 5.4.**

Skill 12.6 Applying principles of electrostatics to determine electrical potential

Electric potential is simply the **potential energy** per unit of charge. Given this definition, it is clear that electric potential must be measured in joules per coulomb and this unit is known as a volt (J/C=V).

Within an electric field there are typically differences in potential energy. This **potential difference** may be referred to as **voltage**. The difference in electrical potential between two points is the amount of work needed to move a unit charge from the first point to the second point.

Stated mathematically, this is:

$$V = \frac{W}{Q}$$

where V= the potential difference
W= the work done to move the charge
Q= the charge

We know from mechanics, however, that work is simply force applied over a certain distance. We can combine this with Coulomb's law to find the work done between two charges distance r apart.

$$W = F \cdot r = k\frac{Q_1 Q_2}{r^2} \cdot r = k\frac{Q_1 Q_2}{r}$$

Now we can simply substitute this back into the equation above for electric potential:

$$V_2 = \frac{W}{Q_2} = \frac{k\frac{Q_1 Q_2}{r}}{Q_2} = k\frac{Q_1}{r}$$

Let's examine a sample problem involving electrical potential.
<u>Problem</u>: What is the electric potential at point A due to the 2 shown charges? If a charge of +2.0 C were infinitely far away, how much work would be required to bring it to point A?

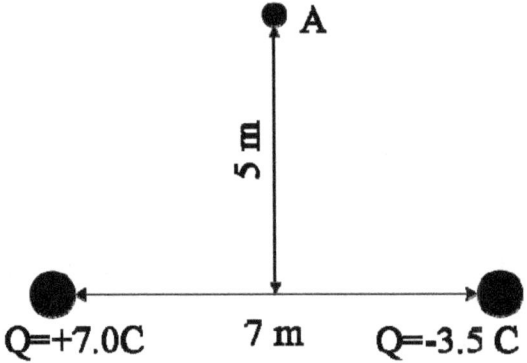

<u>Solution</u>: To determine the electric potential at point A, we simple find and add the potential from the two charges (this is the principle of superposition). From the diagram, we can assume that A is equidistant from each charge. Using the Pythagorean theorem, we determine this distance to be 6.1 m.

$$V = \frac{kq}{r} = k\left(\frac{7.0C}{6.1m} + \frac{-3.5C}{6.1m}\right) = 9 \times 10^9 \frac{N \cdot m^2}{C^2}\left(0.57\frac{C}{m}\right) = 5.13 \times 10^9 V$$

Now, let's consider bringing the charged particle to point A. We assume that electric potential of these particle is initially zero because it is infinitely far away. Since now know the potential at point A, we can calculate the work necessary to bring the particle from V=0, i.e. the potential energy of the charge in the electrical field:

$$W = VQ = (5.13 \times 10^9) \times 2J = 10.26 \times 10^9 J$$

The large results for potential and work make it apparent how large the unit coulomb is. For this reason, most problems deal in microcoulombs (μC).

Skill 12.7 Applying the relationship among capacitance, charge, and potential difference

Capacitance (C) is a measure of the stored electric charge per unit electric potential. The mathematical definition is:

$$C = \frac{Q}{V}$$

It follows from the definition above that the units of capacitance are coulombs per volt, a unit known as a farad ($F=C/V$). In circuits, devices called parallel plate capacitors are formed by two closely spaced conductors. The function of capacitors is to store electrical energy. When a voltage is applied, electrical charges build up in both the conductors (typically referred to as plates). These charges on the two plates have equal magnitude but opposite sign. The capacitance of a capacitor is a function of the distance d between the two plates and the area A of the plates:

$$C \approx \frac{\varepsilon A}{d}; A \gg d^2$$

Capacitance also depends on the permittivity of the non-conducting matter between the plates of the capacitor. This matter may be only air or almost any other non-conducting material and is referred to as a **dielectric**. The permittivity of empty space ε_0 is roughly equivalent to that for air, ε_{air}=8.854x10^{-12} C^2/N•m^2. For other materials, the dielectric constant, κ, is the permittivity of the material in relation to air ($\kappa=\varepsilon/\varepsilon_{air}$). The make-up of the dielectric is critical to the capacitor's function because it determines the maximum energy that can be stored by the capacitor. This is because an overly strong electric field will eventually destroy the dielectric.

TEACHER CERTIFICATION STUDY GUIDE

In summary, a capacitor is "charged" as electrical energy is delivered to it and opposite charges accumulate on the two plates. The two plates generate electric fields and a voltage develops across the dielectric. The energy stored in the capacitor, then, is equal to the amount of work necessary to create this voltage. The mathematical statement of this is:

$$E_{stored} = \frac{1}{2}CV^2 = \frac{1}{2}\frac{Q^2}{C} = \frac{1}{2}VQ$$

The work per unit volume or the **electric field energy density** within a capacitor can be shown to be $\eta = \frac{1}{2}\varepsilon E^2$. This result is generally valid for the energy per unit volume of any electrostatic field, not only for a constant field within a capacitor.

Problem: Imagine that a parallel plate capacitor has an area of 10.00 cm^2 and a capacitance of 4.50 pF. The capacitor is connected to a 12.0 V battery. The capacitor is completely charged and then the battery is removed. What is the separation of the plates in the capacitor? How much energy is stored between the plates? We've assumed that this capacitor initially had no dielectric (i.e., only air between the plates) but now imagine it has a Mylar dielectric that fully fills the space. What will the new capacitance be? (for Mylar, □=3.5)

Solution: To determine the separation of the plates, we use our equation for a capacitor:

$$C = \frac{\varepsilon_0 A}{d}$$

We can simply solve for d and plug in our values:

$$d = \varepsilon_0 \frac{A}{C} = \left(8.854 \times 10^{-12} \frac{C}{N \cdot m^2}\right) \frac{10 \times 10^{-4} m^2}{4.5 \times 10^{-12} F} = 1.97 \times 10^{-3} m = 1.97 mm$$

Similarly, to find stored energy, we simply employ the equation above:

$$E_{stored} = \frac{1}{2}QV$$

But we don't yet know the charge Q, so we must first find it from the definition of capacitance:

$$C = \frac{Q}{V}$$

$$Q = CV = (4.5 \times 10^{-12}) \times (12V) = 5.4 \times 10^{-11} C$$

PHYSICS

Now we can find the stored energy:

$$E_{stored} = \frac{1}{2}QV = \frac{1}{2}(5.4 \times 10^{-11}C)(12V) = 3.24 \times 10^{-10} J$$

To find the capacitance with a Mylar dielectric, we again use the equation for capacitance of a parallel plate capacitor. Note that the new capacitance can be found by multiplying the original capacitance by κ:

$$C = \frac{\kappa_{Mylar}\varepsilon_0 A}{d} = \kappa_{Mylar} C_0 = 3.5 \times 4.5 pF = 15.75 pF$$

TEACHER CERTIFICATION STUDY GUIDE

COMPETENCY 13.0 UNDERSTAND AND ANALYZE CHARACTERISTICS OF DC CIRCUITS

Skill 13.1 Analyzing a DC circuit in terms of conservation of energy and conservation of charge (i.e., Kirchhoff's laws)

Kirchoff's Laws are a pair of laws that apply to conservation of charge and energy in circuits and were developed by Gustav Kirchoff.

Kirchoff's Current Law: At any point in a circuit where charge density is constant, the sum of currents flowing toward the point must be equal to the sum of currents flowing away from that point.

Kirchoff's Voltage Law: The sum of the electrical potential differences around a circuit must be zero.

While these statements may seem rather simple, they can be very useful in analyzing DC circuits, those involving constant circuit voltages and currents.

Problem:
The circuit diagram at right shows three resistors connected to a battery in series. A current of 1.0 A is generated by the battery. The potential drop across R_1, R_2, and R_3 are 5V, 6V, and 10V. What is the total voltage supplied by the battery?

Solution:
Kirchoff's Voltage Law tells us that the total voltage supplied by the battery must be equal to the total voltage drop across the circuit. Therefore:

$$V_{battery} = V_{R_1} + V_{R_2} + V_{R_3} = 5V + 6V + 10V = 21V$$

Problem:
The circuit diagram at right shows three resistors wired in parallel with a 12V battery. The resistances of R_1, R_2, and R_3 are 4 Ω, 5 Ω, and 6 Ω, respectively. What is the total current?

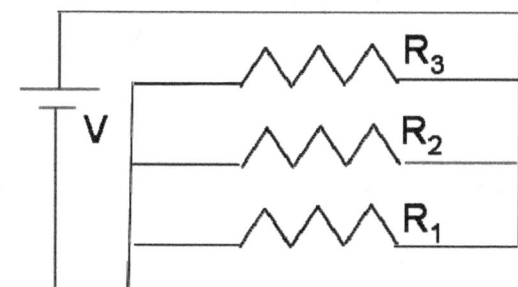

Solution:
This is a more complicated problem. Because the resistors are wired in parallel, we know that the voltage entering each resistor must be the same and equal to that supplied by the battery. We can combine this knowledge with Ohm's Law (see next section) to determine the current across each resistor:

$$I_1 = \frac{V_1}{R_1} = \frac{12V}{4\Omega} = 3A$$

$$I_2 = \frac{V_2}{R_2} = \frac{12V}{5\Omega} = 2.4A$$

$$I_3 = \frac{V_3}{R_3} = \frac{12V}{6\Omega} = 2A$$

Finally, we use Kirchoff's Current Law to find the total current:

$$I = I_1 + I_2 + I_3 = 3A + 2.4A + 2A = 7.4A$$

Skill 13.2 Analyzing the interrelationships among potential difference, resistance, and current

Within an electric field there are typically differences in potential energy. This potential difference is usually referred to as voltage. Since electric potential is the potential energy per unit of charge, it is clear that electric potential must be measured in joules per coulomb and this unit is known as a volt. Electrical current, however, is the rate at which charge flows past a given point. Current is measured in amperes which is equivalent to coulombs/ second.

When electrical charge flows through a conductor, both current and a potential difference will be present. It is logical that a greater potential difference will lead to a faster flow of charge and, therefore, a higher current:

$$V \alpha I$$

As in any other proportionality relationship, we can simply add a proportionality constant:

$$V = kI$$

In the case of electricity, the proportionality constant is a measure of the material's resistance to current. This quantity is known as resistance and is a constant for a given material at a given temperature regardless of voltage or current. Restating our equation:

$$V = RI$$

This expression is Ohm's Law. Note that resistance must have units of $\frac{J \cdot s}{C^2}$ and this is known as ohm (Ω).

We can draw a rough analogy between this electrical phenomenon and simple mechanics. Imagine a block sliding down an inclined plane under the force of gravity. The speed of the block as it travels from one point to another is directly proportional to the difference in gravitational potential energy between those points. However, the block's speed can be slowed by the friction between the block and plane. Thus friction is similar to resistance, voltage similar to the difference in gravitational potential energy, and current similar to speed. The voltage drop, like gravity, provides the driving force while resistance impedes the flow.

Skill 13.3   Analyzing factors that affect resistance

Conductors are those materials which allow for the free passage of electrical current. However, all materials exhibit a certain opposition to the movement of electrons. This opposition is known as resistivity (ρ). Resistivity is determined experimentally by measuring the resistance of a uniformly shaped sample of the material and applying the following equation:

$$\rho = R\frac{A}{l}$$

where ρ = static resistivity of the material
R = electrical resistance of the material sample
A = cross-sectional area of the material sample
L = length of the material sample

The temperature at which these measurements are taken is important as it has been shown that resistivity is a function of temperature. For conductors, resistivity increases with increasing temperature and decreases with decreasing temperature. At extremely low temperatures resistivity assumes a low and constant value known as residual resistivity (ρ_0). Residual resistivity is a function of the type and purity of the conductor.

The following equation allows us to calculate the resistivity ρ of a material at any temperature given the resistivity at a reference temperature, in this case at $20^\circ C$:

$$\rho = \rho_{20}[1 + a(t - 20)]$$

where ρ_{20} = resistivity at $20^\circ C$

a = proportionality constant characteristic of the material

t = temperature in Celcius

Problem: The tungsten filament in a certain light bulb is a wire 8 μm in diameter and 10 mm long. Given that, for tungsten, $\rho_{20} = 5.5 \times 10^{-8}$ Ω·m and $a = 4.5 \times 10^{-3} K^{-1}$, what will the resistance of the filament be at 45°C?

Solution: First we must find the resistivity of the tungsten at 45°C:

$$\rho = 5.5 \times 10^{-8}(1 + 4.5 \times 10^{-3}(45 - 20)) = 6.1 \times 10^{-8} \Omega \cdot m$$

Now we can rearrange the equation defining resistivity and solve for the resistance of the filament:

$$R = \rho\frac{l}{A} = 6.1 \times 10^{-8} \times 0.01 / (\pi(4 \times 10^{-6})^2) = 12.1\Omega$$

Skill 13.4 Solving problems involving Ohm's law

Ohm's Law is the most important tool we posses to analyze electrical circuits. Ohm's Law states that the current passing through a conductor is directly proportional to the voltage drop and inversely proportional to the resistance of the conductor. Stated mathematically, this is:

$$V=IR$$

Problem:
The circuit diagram at right shows three resistors connected to a battery in series. A current of 1.0A flows through the circuit in the direction shown. It is known that the equivalent resistance of this circuit is 25 Ω. What is the total voltage supplied by the battery?

Solution:

To determine the battery's voltage, we simply apply Ohm's Law:

$$V = IR = 1.0A \times 25\Omega = 25V$$

Skill 13.5 Describing how to use various meters to measure properties of an electric circuit

Electrical meters function by utilizing the following familiar equations:

Across a resistor (Resistor R):

$$V_R = IR_R$$

Across a capacitor (Capacitor C):

$$V_C = IX_C$$

Across an inductor (Inductor L):

$$V_L = IX_L$$

Where V=voltage, I=current, R=resistance, X=reactance.

Ammeter: An ammeter placed in series in a circuit measures the current through the circuit. An ammeter typically has a very small resistance so that the current in the circuit is not changed too much by insertion of the ammeter.

Voltmeter: A voltmeter is used to measure potential difference. The potential difference across a resistor is measured by a voltmeter placed in parallel across it. An ideal voltmeter has very high resistance so that it does not appreciably alter circuit resistance and therefore the voltage drop it is measuring.

Galvanometer: A galvanometer is a device that measures current and is a component of an ammeter or a voltmeter. A typical galvanometer consists of a coil of wire, an indicator and a scale that is designed to be proportional to the current in the galvanometer. The principle that a current-carrying wire experiences a force in a magnetic field is used in the construction of a galvanometer. In order to create a voltmeter from a galvanometer, resistors are added in series to it. To build an ammeter using a galvanometer, a small resistance known as a **shunt resistor** is placed in parallel with it.

Potentiometer: A potentiometer is a variable resistance device in which the user can vary the resistance to control the current and voltage applied to a circuit. Since the potentiometer can be used to control what fraction of the emf of a battery is applied to a circuit, it is also known as a voltage divider. It can be used to measure an unknown voltage by comparing it with a known value.

Multimeter: A common electrical meter, typically known as a multimeter, is capable of measuring voltage, resistance, and current. Many of these devices can also measure capacitance (farads), frequency (hertz), duty cycle (a percentage), temperature (degrees), conductance (siemens), and inductance (henrys).

Skill 13.6   Interpreting schematic diagrams of electric circuits

The two most important elements in simple circuits are resistors and capacitors. Often resistors and capacitors are used together in series or parallel. Two components are in series if one end of the first element is connected to one end of the second component. The components are in parallel if both ends of one element are connected to the corresponding ends of another. A series circuit has a single path for current flow through all of its elements. A parallel circuit is one that requires more than one path for current flow in order to reach all of the circuit elements.

Below is a diagram demonstrating a simple circuit with resistors in parallel (on right) and in series (on left). Note the symbols used for a battery (noted V) and the resistors (noted R).

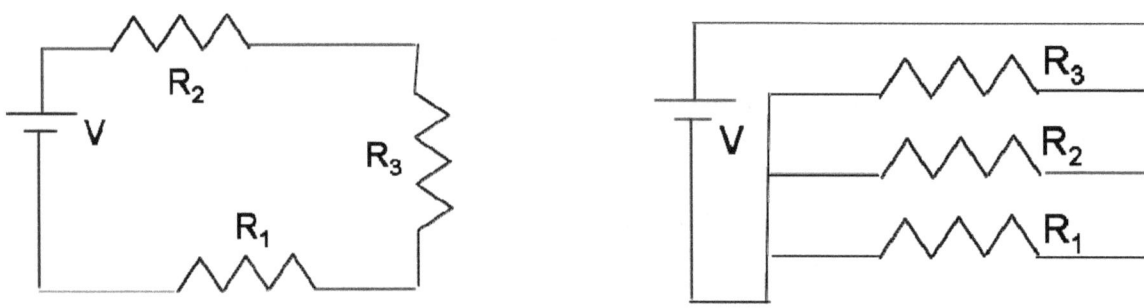

Thus, when the resistors are placed in series, the current through each one will be the same. When they are placed in parallel, the voltage through each one will be the same. To understand basic circuitry, it is important to master the rules by which the equivalent resistance (R_{eq}) or capacitance (C_{eq}) can be calculated from a number of resistors or capacitors:

Resistors in parallel: $\dfrac{1}{R_{eq}} = \dfrac{1}{R_1} + \dfrac{1}{R_2} + \cdots + \dfrac{1}{R_n}$

Resistors in series: $R_{eq} = R_1 + R_2 + \cdots + R_n$

Capacitors in parallel: $C_{eq} = C_1 + C_2 + \cdots + C_n$

Capacitors in series: $\dfrac{1}{C_{eq}} = \dfrac{1}{C_1} + \dfrac{1}{C_2} + \cdots + \dfrac{1}{C_n}$

Skill 13.7 Analyzing work, energy, and power in DC circuits

Electrical power is a measure of how much energy is expended or how much work can be done by an electrical current and has units of watts (W). To determine electrical power, we simply use Joule's law:

$$P = IV$$

where P=power
I=current
V=voltage

If we combine this with Ohm's law (V=IR), we generate two new equations that are useful for finding the amount of power dissipated by a resistor:

$$P = I^2 R$$

$$P = \frac{V^2}{R}$$

Problem: How much power is dissipated by a 1 kΩ resistor with a 50V voltage drop through it?

Solution:

$$P = \frac{V^2}{R} = \frac{(50V)^2}{1000\Omega} = 2.5W$$

Skill 13.8 Describing the properties of conductors, semiconductors, superconductors (e.g., critical temperature, Meissner effect)

Semiconductors, conductors and superconductors are differentiated by how "easily" current can flow in the presence of an applied electric field. Dielectrics (insulators) conduct little or no current in the presence of an applied electric field, as the charged components of the material (for example, atoms and their associated electrons) are tightly bound and are not free to move within the material. In the case of materials with some amount of conductivity, so-called valence (higher energy level) electrons are only loosely bound and may move among positive charge centers (i.e., atoms or ions).

Conductors, such as metals like aluminum and iron, have numerous valence electrons distributed among the atoms that compose the material. These electrons can move freely, especially under the influence of an applied electric field. In the absence of any applied field, these electrons move randomly, resulting in no net current. If a static electric field is applied, the mobile electrons reorient themselves to minimize the energy of the system and create an equipotential on the surface of the metal (in the ideal case of infinite conductivity). The highly mobile charge carriers do not allow a net field inside the metal; thus conductors can "shield" electromagnetic fields.

Semiconductors, such as silicon and germanium, are materials that can neither be described as conductors, nor as insulators. **A certain number of charge carriers are mobile in the semiconductor**, but this number is nowhere near the free charge populations of conductors such as metals. "Doping" of a semiconductor by adding so-called donor atoms or acceptor atoms to the intrinsic (or pure) semiconductor can increase the conductivity of the material. Furthermore, combination of a number of differently doped semiconductors (such as donor-doped silicon and acceptor-doped silicon) can produce a device with beneficial electrical characteristics (such as the diode). The conductivity of such devices can be controlled by applying voltages across specific portions of the device.

A useful way to visualize the relationship of conductors and semiconductors is by way of energy band diagrams. For purposes of comparison, an example of an insulator is included here as well.

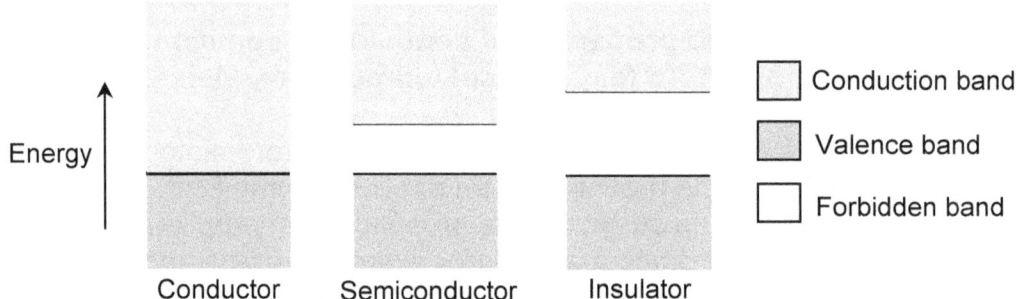

According to the theory of energy bands, as derived from quantum mechanics, electrons in the ground state reside in the valence band and are bound to their associated atoms or molecules. If the electrons gain sufficient energy, such as through heat, they can jump across the forbidden band (in which no electrons may exist) to the conduction band, thus becoming free electrons that can form a current in the presence of an applied field. For conductors, the valence band and conduction band meet or overlap, allowing electrons to easily jump to unoccupied conduction states. Thus, conductors have an abundance of free electrons. It is noteworthy that only two bands are shown in the diagram above, but that an infinite number of bands may exist at higher energies. A more general statement of the difference in band structures is that, for insulators and semiconductors, the valence electrons fill up all the states in a particular band, leaving a gap between the highest energy valence electrons and the next available band. The difference between these two types of materials is simply a matter of the "size" of the forbidden band. For conductors, the band that contains the highest energy electrons has additional available states.

A nearly ideal conducting material is a superconductor. As a material increases in temperature, increased vibrational motion of the atoms or molecules leads to decreased charge carrier mobility and decreased conductivity. In the case of semiconductors, the increase in free carrier population outweighs the loss in mobility of the charge carriers, meaning that the semiconductor increases in conductivity as temperature increases. Superconductors, on the other hand, reach their peak conductivity at extremely low temperatures (although there are currently numerous efforts to achieve superconductivity at higher and higher temperatures, with room temperature or higher being the ultimate goal). The critical temperature of the material is the temperature at which superconducting properties emerge. At this temperature, the material has a nearly infinite conductivity and maintains an almost perfect equipotential across its surface when in the presence of a static electric field. Inside a superconductor, the electric field is virtually zero at all times. As a result, the time derivative of the electric flux density is zero, and, by Maxwell's equations, the magnetic flux density must likewise be zero.

$$\nabla \times \mathbf{H} = \frac{\partial \mathbf{D}}{\partial t} + \mathbf{J}$$

Since the electric field is also zero, the current density **J** inside the superconductor must also be zero (or very nearly so). **This elimination of the magnetic flux density inside a superconductor is called the Meissner effect.**

Skill 13.9 Describing the function of a solid state device in an electric circuit

Prior to the 1960s most electronic apparatuses relied on either vacuum tubes or mechanical devices, typically relays. As semiconductors became increasingly available, however, vacuum tubes were replaced with solid state devices. Vacuum tubes conduct electrons through a heated vacuum but semiconductors allow electrons to flow through them while still solid. Solid state has come to mean any circuit that does not contain the aforementioned vacuum tubes and, in short, operates with "no moving parts". The lack of these elements make solid state devices more resistant to physical stressors, such as vibration, and more durable in general, since they are less susceptible to wear. A familiar example of this are the solid state "flash cards" that are popular for data storage. Previously, hard disk players were primarily used for a similar function but their moving parts make them of limited durability and less practical to transport.

Semiconductors are extremely useful because their conductive properties can be controlled. This is done by "doping" or introducing impurities. The impurity or dopant is used to introduce extra electrons or extra free orbital space that can be filled with electrons. This allows much freer movement of electron and, therefore, flow of current through the material. Semiconductors doped to contain extra electrons are known as N-type, while those doped to contain extra "holes" are known a P-type. Typically, either type of semiconductor can be made from the same base material. For instance, silicon can be doped with boron to create a P-type semiconductor or with phosphorus to create an N-type semiconductor. Two of the most common and important semiconductor devices today are diodes and transistors.

Diodes: Diodes restrict the direction of current (electron) flow in a given direction. Somewhat analagous to a check valve, they allow the electrical current to flow in one direction, but not in the other. Thus diodes are found in almost all circuits where single direction current flow is required. While early diodes were made from vacuum tubes, today most are made from semiconductors. This is often done by desiging a diode with P-type semiconduting material on one side and N-type on the other. The current flows into the P-type side and on to the N-type side, but cannot flow in the opposite direction. Alternatively, the diode can be made from a conducting metal and a semiconductor and function in a similar fashion.

Transistors: Just like the vaccum tubes they replaced, transistors control current flow. They serve a variety of functions in circtuits and can act as amplifiers, switches, voltage regulators, signal modulators, or oscillators. They are perhaps the most important building block of modern circuitry and electronics. As in solid state diodes, the properties of semiconductors are exploited in transistors to control the flow of current. However, where the diode is analagous to a check valve, a transistor function more like a tap on a sink and is able to control the rate of current flow or eliminate it all together. Today, most transistors are either bipolar junction transistors (BJT) or field effect transistors (FET). Both accomplish control of current flow, but it is done by applying current in the BJTs and voltage in the FETs.

COMPETENCY 14.0 UNDERSTAND MAGNETS, ELECTROMAGNETS, AND MAGNETIC FIELDS; THE EFFECTS OF MAGNETIC FIELDS ON MOVING ELECTRIC CHARGES; AND THE APPLICATIONS OF ELECTROMAGNETISM

Skill 14.1 Describing magnetic fields as the result of moving electric charges

Conductors through which electrical currents travel will produce magnetic fields: The magnetic field dB induced at a distance r by an element of current Idl flowing through a wire element of length dl is given by the **Biot-Savart** law

$$dB = \frac{\mu_0}{4\pi} \frac{Idl \times \hat{r}}{r^2}$$

where μ_0 is a constant known as the permeability of free space and $\hat{r}$ is the unit vector pointing from the current element to the point where the magnetic field is calculated.

An alternate statement of this law is **Ampere's law** according to which the line integral of $B.dl$ around any closed path enclosing a steady current I is given by

$$\oint_C B \cdot dl = \mu_0 I$$

The basis of this phenomenon is the same no matter what the shape of the conductor, but we will consider three common situations:

Straight Wire

Around a current-carrying straight wire, the magnetic field lines form concentric circles around the wire. The direction of the magnetic field is given by the right-hand rule: When the thumb of the right hand points in the direction of the current, the fingers curl around the wire in the direction of the magnetic field. Note the direction of the current and magnetic field in the diagram.

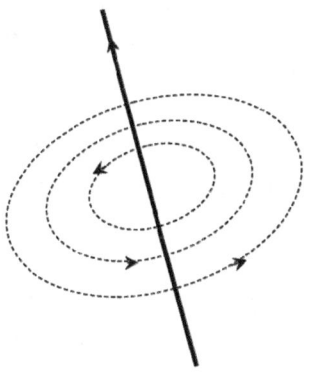

To find the magnetic field of an infinitely long (allowing us to disregarding end effects) we apply Ampere's Law to a circular path at a distance r around the wire:

$$B = \frac{\mu_0 I}{2\pi r}$$

where μ_0=the permeability of free space ($4\pi \times 10^{-7}$ T·m/A)
I=current
r=distance from the wire

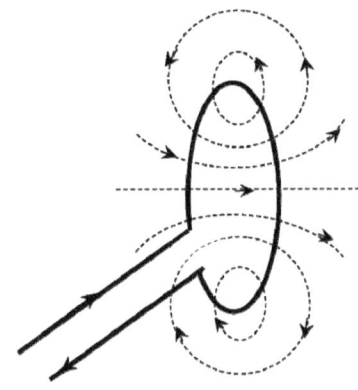

Loops
Like the straight wire from which it's been made, a looped wire has magnetic field lines that form concentric circles with direction following the right-hand rule. However, the field are additive in the center of the loop creating a field like the one shown. The magnetic field of a loop is found similarly to that for a straight wire.

In the center of the loop, the magnetic field is:

$$B = \frac{\mu_0 I}{2r}$$

Solenoids
A solenoid is essentially a coil of conduction wire wrapped around a central object. This means it is a series of loops and the magnetic field is similarly a sum of the fields that would form around several loops, as shown.

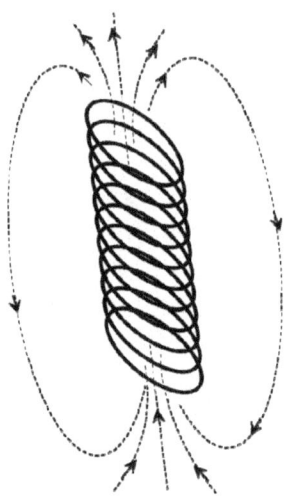

The magnetic field of a solenoid can be found as with the following equation:

$$B = \mu_0 n I$$

In this equation, n is turn density, which is simply the number of turns divided by the length of the solenoid.

Skill 14.2 Applying the domain theory to the magnetization of ferromagnetic materials

A ferromagnetic material can be either fully magnetized (saturation) or completely unmagnetized. Ferromagnetic domain theory presents an explanation of this phenomenon by viewing the material as being divided into a number of small or microscopic "domains" each of which has its own associated magnetic dipole moment that is aligned in a particular direction. This domain magnetic moment results from the alignment of the moments of the individual atoms in the domain. The overall magnetization of the macroscopic material is the sum of the contributions of all the domains contained in the material.

When the magnetic moments of the individual domains are randomly directed, no net magnetization (**M**) occurs. Full magnetization (saturation) occurs when the magnetic moments of all the domains are aligned in the same direction. A range of magnetizations for the macroscopic material, between the extremes of zero and saturation, can be achieved depending on the degree of alignment of the domains in the material.

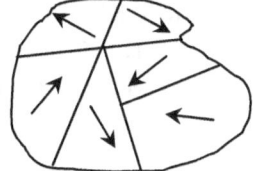
Random alignment
No net magnetization

Partial alignment
Partial magnetization

Full alignment
Saturation

Ferromagnetic domains can be as large as a few millimeters or as small as a micrometer and they can vary in shape. If a magnetic field (**H**) is applied to the material, a net magnetization is achieved due to either a realignment of the domains in the direction of the applied field or due to a change in the size of the domains favorable to a net magnetization in the direction of the applied field.

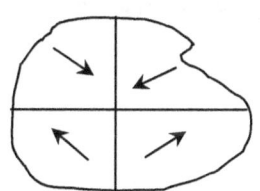
Random alignment
No net magnetization

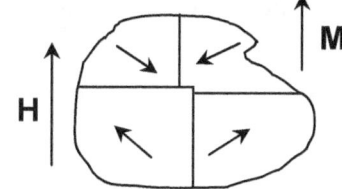
Magnetization by movement
of domain boundaries

Magnetization by
alignment of domains

The level of magnetization in the material depends on the strength of the applied field and can range from a small amount all the way to saturation. When the applied field is removed, some magnetization may remain however. Residual magnetization implies that the present state of the material, magnetically, depends on its past magnetic history (hysteresis). A so-called hysteresis loop describes the magnetization of the material in the presence of an applied field.

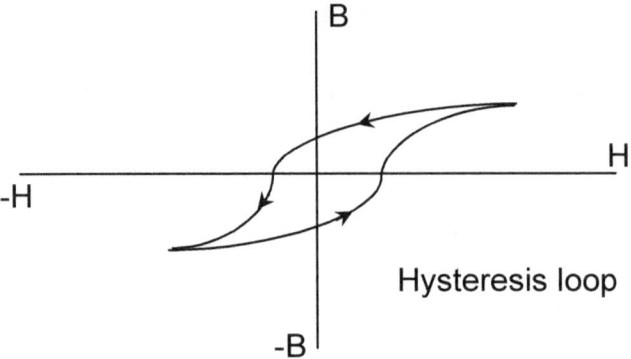

Hysteresis loop

The above depiction is for one possible hysteresis loop for some particular material. There are, in fact, an infinite number of loops, depending on the range of applied field strengths. The shape depends on both present and past values of the applied field as well as the tendency of the ferromagnetic domains to return to their original (possibly random) alignment when the applied field is decreased or eliminated.

Skill 14.3 Determining the orientation and magnitude of a magnetic field

Magnetic field lines are a good way to visualize a magnetic field. The distance between magnetic fields lines indicates the strength of the magnetic field such that the lines are closer together near the poles of the magnets where the magnetic field is the strongest. The lines spread out above and below the middle of the magnet, as the field is weakest at those points furthest from the two poles. The SI unit for magnetic field known as magnetic induction is Tesla(T) given by 1T = 1 N.s/(C.m) = 1 N/(A.m). Magnetic fields are often expressed in the smaller unit Gauss (G) (1 T = 10,000 G). Magnetic field lines always point from the north pole of a magnet to the south pole.

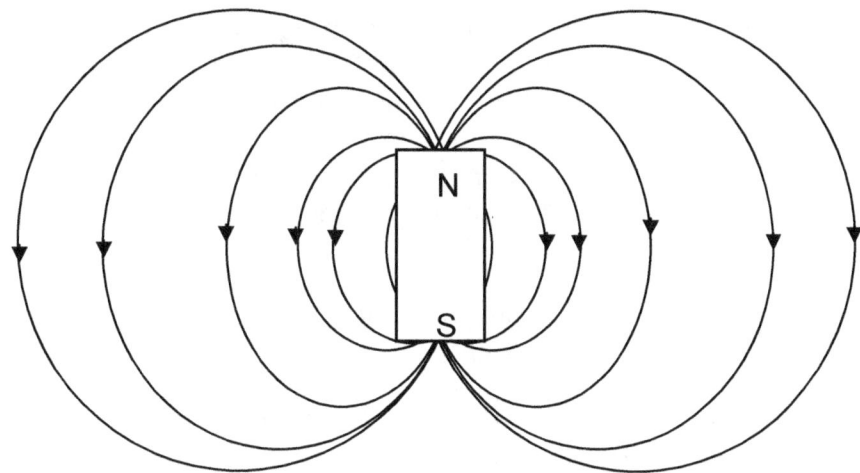

Magnetic field lines can be plotted with a magnetized needle that is free to turn in 3 dimensions. Usually a compass needle is used in demonstrations. The direction tangent to the magnetic field line is the direction the compass needle will point in a magnetic field. Iron filings spread on a flat surface or magnetic field viewing film which contains a slurry of iron filings are another way to see magnetic field lines.

When two magnets are placed close to one another magnetic field lines form between the north and south pole of each magnet individually and the field line also interact between the two magnets. The magnetic field in the area between the two magnets will depend on how the magnets are oriented with respect to one another. If the north pole of one magnet is placed near the south pole of the other magnet, for instance, you can see field lines starting at that north pole and ending at the south pole of the other magnet. If additional magnets and metal objects are placed in the arrangement the interactions become more complicated and various patterns are formed by the magnetic field lines.

Skill 14.4 Determining the magnitude and direction of the force on a charge or charges moving in a magnetic field

The magnetic force exerted on a charge moving in a magnetic field depends on the size and velocity of the charge as well as the magnitude of the magnetic field. One important fact to remember is that only the velocity of the charge in a direction perpendicular to the magnetic field will affect the force exerted. Therefore, a charge moving parallel to the magnetic field will have no force acting upon it whereas a charge will feel the greatest force when moving perpendicular to the magnetic field.

The direction of the magnetic force is always at a right angle to the plane formed by the velocity vector v and the magnetic field B and is given by applying the right hand rule - if the fingers of the right hand are curled in a way that seems to rotate the v vector into the B vector, the thumb points in the direction of the force. The magnitude of the force is equal to the cross product of the velocity of the charge with the magnetic field multiplied by the magnitude of the charge.

$$F=q (v \times B) \quad or \quad F=q v B\sin(\theta)$$

Where θ is the angle formed between the vectors of velocity of the charge and direction of magnetic field.

Problem: Assuming we have a particle of 1 x 10⁻⁶ kg that has a charge of -8 coulombs that is moving perpendicular to a magnetic field in a clockwise direction on a circular path with a radius of 2 m and a speed of 2000 m/s, let's determine the magnitude and direction of the magnetic field acting upon it.

Solution: We know the mass, charge, speed, and path radius of the charged particle. Combining the equation above with the equation for centripetal force we get

$$qvB = \frac{mv^2}{r} \quad or \quad B = \frac{mv}{qr}$$

Thus B= (1 x 10⁻⁶ kg) (2000m/s) / (-8 C)(2 m) = 1.25 x 10⁻⁴ Tesla

Since the particle is moving in a clockwise direction, we use the right hand rule and point our fingers clockwise along a circular path in the plane of the paper while pointing the thumb towards the center in the direction of the centripetal force. This requires the fingers to curl in a way that indicates that the magnetic field is pointing out of the page. However, since the particle has a negative charge we must reverse the final direction of the magnetic field into the page.

A mass spectrometer measures the mass to charge ratio of ions using a setup similar to the one described above. m/q is determined by measuring the path radius of particles of known velocity moving in a known magnetic field.

Skill 14.5 Analyzing the behavior of a current-carrying wire in a magnetic field

The **force on a current-carrying conductor** in a magnetic field is the sum of the forces on the moving charged particles that create the current. For a current I flowing in a straight wire segment of length l in a magnetic field B, this force is given by

$$\mathbf{F} = I\mathbf{l} \times \mathbf{B}$$

where **l** is a vector of magnitude l and direction in the direction of the current.

When a current-carrying loop is placed in a magnetic field, the net force on it is zero since the forces on the different parts of the loop act in different directions and cancel each other out. There is, however, a net torque on the loop that tends to rotate it so that the area of the loop is perpendicular to the magnetic field. For a current I flowing in a loop of area A, this torque is given by

$$\tau = IA\hat{n} \times \mathbf{B}$$

where $\hat{n}$ is the unit vector perpendicular to the plane of the loop.

Skill 14.6 Analyzing factors that affect the strength of an electromagnet

The movement of electric charges (i.e., a current density **J**) results in a magnetic field **H**, as described by the Maxwell equation based on Ampere's law:

$$\nabla \times \mathbf{H}(r,t) = \frac{\partial \mathbf{D}(r,t)}{\partial t} + \mathbf{J}(r,t)$$

For the case where the time derivative of the electric flux density **D** is zero, this reduces to the simpler Ampere's law:

$$\nabla \times \mathbf{H}(r,t) = \mathbf{J}(r,t)$$

Electromagnets take advantage of this relationship between the current and the magnetic field by, for example, coiling a current-carrying wire (a solenoid), sometimes around a ferromagnetic or paramagnetic material. This creates a magnetic dipole with a strength that varies depending on a number of factors.

The direction of the dipole relative to the direction of current flow is determined by the right hand rule: the fingers curl in the direction of the current, and the thumb points in the (north) direction of the dipole.

The strength of the electromagnet (i.e., the induced magnetic dipole) can be increased by adding to the number of turns (or loops) of current-carrying wire. The principle of superposition applies here and the induced magnetic field varies proportionally according to the number of turns. Also, by varying the current, the strength of the electromagnet can be increased or decreased. Imperfections in the solenoid, such as imperfectly packed wire loops, which can cause magnetic flux leakage, can adversely affect the strength of the electromagnet. Also, the spatial extent of the solenoid (or other form of electromagnet, such as a toroid) must be considered if exact numerical calculations are pursued.

Another critical factor that affects the electromagnet is the material used inside the coil; specifically, the magnetic properties of the material. Magnetizable materials with permeabilities (μ) greater than unity can be used to increase the magnitude of the induced magnetic dipole. Ferromagnetic materials, for example, can have very large permeabilities. One such example is iron. A ferromagnetic core is magnetized when the magnetic field induced by the current-carrying wire is applied resulting in a stronger magnetic dipole than would have been produced if a non-magnetic material (such as air) had been used. The ability of a magnetic material core to increase the strength of the electromagnet is limited, however. In the case of a ferromagnetic material, once all the magnetic domains have been aligned with the magnetic field of the electromagnet, saturation has been reached and the material cannot be magnetized further.

Skill 14.7 Analyzing the use of electromagnetism in technology (e.g., motors, generators, meters, magnetic media)

Electromagnetism is the foundation for a vast number of modern technologies ranging from computers to communications equipment. More mundane technologies such as motors and generators are also based upon the principles of electromagnetism. The particular understanding of electrodynamics can either be in terms of quantum mechanics (quantum electrodynamics) or classical electrodynamics, depending on the type of phenomenon being analyzed. In classical electrodynamics, which is a sufficient approximation for most situations, the electric field, resulting from electric charge, and the magnetic field, resulting from moving charges, are the parameters of interest and are related through Maxwell's equations.

Motors
Electric motors are found in many common appliances such as fans and washing machines. The operation of a motor is based on the principle that a magnetic field exerts a force on a current carrying conductor. This force is essentially due to the fact that the current carrying conductor itself generates a magnetic field; the basic principle that governs the behavior of an electromagnet. In a motor, this idea is used to convert **electrical energy into mechanical energy**, most commonly rotational energy. Thus the components of the simplest motors must include a strong magnet and a current-carrying coil placed in the magnetic field in such a way that the force on it causes it to rotate.

A typical motor is composed of a stationary portion, called the stator, and a rotating (or moving) portion, called the rotor. Coils of wire that serve as electromagnets are wound on the armature, which can be either the stator or the rotor, and are powered by an electric source. Motors use electric current to generate a magnetic field around an electromagnet, which results in a rotational force due to the presence of an external magnetic field (either from permanent magnets or electromagnets). The designs of various motors can differ dramatically, but the general principles of electromagnetism that describe their operation are generally the same.

Generators

Generators are in effect "reverse motors". They exploit electromagnetic induction (**see Skill 15.1**) to generate electricity. Thus, if a coil of wire is rotated in a magnetic field, an alternating EMF is produced which allows current to flow. Any number of energy sources can be used to rotate the coil, including combustion, nuclear fission, flowing water or other sources. Therefore, generators are devices that convert **mechanical or other forms of energy into electrical energy**. See **Skill 15.4** for more about generators.

Meters

A number of different types of meters use electromagnetism or are designed to measure certain electromagnetic parameters. For example, older forms of ammeters (galvanometers), when supplied with a current, provided a measurement through the deflection of a spring-loaded needle. A coil connected to the needle acted as an electromagnet which, in the presence of a permanent magnetic field, would be deflected in the same manner as a rotor, as mentioned previously. The strength of the electromagnet, and thus the extent of the deflection, is proportional to the current. Further, the spring limits the deflection in such a manner that a reasonably accurate measurement of the current is provided. For additional information on electrical meters see **Skill 13.5**.

Magnetic Media

Although the cassette tape has fallen out of favor with popular culture, magnetic media are still widely used for information storage. Magnetic strips on the back of credit and identification cards, computer hard drive disks and magnetic tapes (such as those contained in cassettes) are all examples of magnetic media. The principle underlying magnetic media is the sequential magnetization of a region of the medium. A special head is able to detect spatial magnetic fluctuations in the medium which are then converted into an electrical signal. The head is often able to "write" to the medium as well. The electrical signal from the medium can be converted into sound or video, as with the video or audio cassette player, or it can be digitized for use in a computer.

COMPETENCY 15.0 UNDERSTAND AND APPLY THE PRINCIPLES OF ELECTROMAGNETIC INDUCTION AND AC CIRCUITS

Skill 15.1 Analyzing factors that affect the magnitude of an induced electromotive force (EMF)

When the magnetic flux through a coil is changed, a voltage is produced which is known as induced electromagnetic force. Magnetic flux is a term used to describe the number of magnetic fields lines that pass through an area and is described by the equation:

$$\Phi = BA\cos\theta$$

Where Φ is the angle between the magnetic field B, and the normal to the plane of the coil of area A

By changing any of these three inputs, magnetic field, area of coil, or angle between field and coil, the flux will change and an EMF can be induced. The speed at which these changes occur also affects the magnitude of the EMF, as a more rapid transition generates more EMF than a gradual one. This is described by **Faraday's law** of induction:

$$\varepsilon = -N\,\Delta\Phi / \Delta t$$

where ε is emf induced, N is the number of loops in a coil, t is time, and Φ is magnetic flux

The negative sign signifies **Lenz's law** which states that induced emf in a coil acts to oppose any change in magnetic flux. Thus the current flows in a way that creates a magnetic field in the direction opposing the change in flux. See the next section for the right-hand rule that determines the direction of the induced current.

Consider a coil lying flat on the page with a square cross section that is 10 cm by 5 cm. The coil consists of 10 loops and has a magnetic field of 0.5 T passing through it coming out of the page. Let's find the induced EMF when the magnetic field is changed to 0.8 T in 2 seconds.

First, let's find the initial magnetic flux: Φ_i

$\Phi_i = BA\cos\theta = (.5\text{ T})(.05\text{ m})(.1\text{ m})\cos 0° = 0.0025\text{ T m}^2$

And the final magnetic flux: Φ_f

$\Phi_f = BA\cos\theta = (0.8\text{ T})(.05\text{ m})(.1\text{ m})\cos 0° = 0.004\text{ T m}^2$

The induced emf is calculated then by

$\varepsilon = -N\,\Delta\Phi / \Delta t = -10\,(.004\text{ T m}^2 - .0025\text{ T m}^2) / 2\text{ s} = -0.0075\text{ volts}.$

To determine the direction the current flows in the coil we need to apply the right hand rule and Lenz's law. The magnetic flux is being increased out of the page, with your thumb pointing up the fingers are coiling counterclockwise. However, Lenz's law tells us the current will oppose the change in flux so the current in the coil will be flowing clockwise.

Skill 15.2 Applying the appropriate hand rule to determine the direction of an induced current

The negative sign in Faraday's Law (described in the previous section) leads to Lenz's law which states that the induced current must produce a magnetic field that opposes the change in the applied magnetic field. This is simply an expression of the conservation of energy (See **Skill 15.3**).

The right- or left-hand rule applies only in the case of a particular current convention: the right-hand rule is typically applied in the case of the positive current convention (i.e., the direction of flow of positive charge). Since positive and negative current are complementary, however, the rules for negative current simply involve using the left hand instead of the right hand. For simplicity, only the right-hand rule will be discussed explicitly.

The right-hand rule states that if the fingers of the right hand are curled in the direction of the positive current flow, the resulting magnetic flux is in the direction of the extended thumb. According to Lenz's law, when the magnetic flux through the surface bounded by the conductor increases, the induced current, $i_{induced}$, must produce a magnetic flux that is in the opposite direction to that of the applied flux. This is illustrated below.

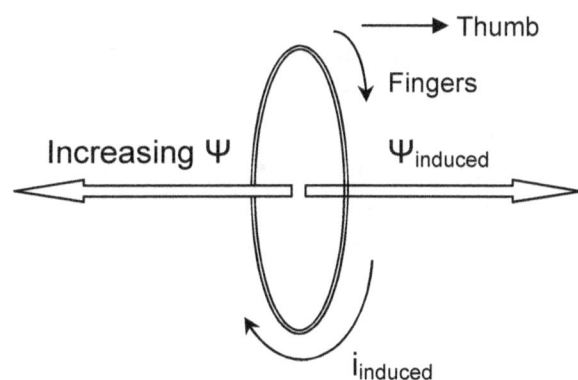

When the applied magnetic flux through the surface decreases, the direction of the induced flux must likewise coincide with the direction of the applied flux.

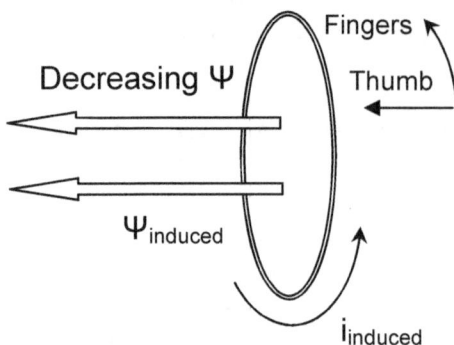

Thus, the induced magnetic flux must oppose the change in the applied flux. The direction of the induced positive current is the same as the direction of the curled fingers of the right hand when the thumb is extended in the direction of the induced magnetic flux. If negative current is of interest (for example, the flow of electrons), then the left-hand must be substituted in place of the right hand; the thumb follows the same rules, but the direction of the curled fingers indicates the direction of negative current flow.

Skill 15.3   Analyzing Lenz's law in terms of conservation of energy

Though it may not be entirely obvious initially, Lenz's law is a consequence of the conservation of energy. We can understand this by considering the physical scenario in which a permanent magnet (grey in diagram) moves through a loop of wire in the direction shown. An electric current will be induced in the wire and its direction will depend on which pole of the magnet enters the loop first. According to Lenz's law, the current will flow in the direction shown in this diagram. The direction of the current shown would be reversed if the magnet were turned so that the south pole entered first or if the magnet traveled in the direction opposite that shown by the arrow.

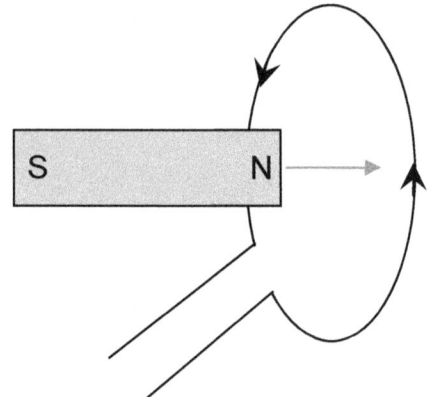

Now, suppose that Lenz's law did not hold and instead the magnet in the situation above induced a current in the direction opposite from that shown. That induced current would then induce a magnetic field directed toward the left of the page. This would, in turn, produce an attractive force between the permanent magnet and the induced magnetic field. A feedback loop would emerge: the permanent magnet would be drawn into the loop more quickly, a greater current induced, a stronger field generated, and the magnet drawn even more quickly into the loop. This would have an effect of increasing the kinetic energy of the magnet while also increasing the rate of energy dissipation of the loop. In short, energy would be created. Since this violation of the law of energy conservation cannot be correct, Lenz's law is supported.

Skill 15.4   Analyzing the functions of transformers and generators

Transformers and generators have critical functions in the production and use of electrical power. A generator is essentially a transducer that converts one form of energy, such as mechanical or heat energy, into electrical energy. It uses electromagnetic induction (**see Skill 15.1**) to generate electricity. The simplest way to implement a generator is to rotate a wire loop in a permanent magnetic field.

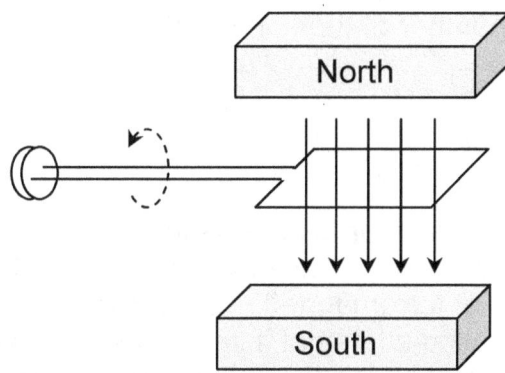

Depending on the strength of the magnetic field and the rate of rotation of the loop, a voltage of a specific amplitude can be produced. Additionally, the rotation of the wire loop means that the magnetic flux is at times increasing and at times decreasing, resulting in a sinusoidal voltage with a frequency equal to the rotation frequency of the loop. Electricity supplied by public utilities, for example, is 120 V at a frequency of 60 Hz. The source of energy for the wire loop rotator (turbine) can be a dammed river or an engine run by combustion, a nuclear reaction or another process. The energy extracted from these sources (either as mechanical or heat energy) is transferred to electrical energy through the use of the generator.

Transformers fulfill another purpose in the context of electrical power by either stepping up or stepping down the voltage (and, concomitantly, stepping down or stepping up the current). They do this by magnetically coupling two circuits together. This allows the transfer of energy between these two circuits without requiring motion. Typically, a transformer consists of a couple of coils and a magnetic core. A changing voltage applied to one coil (the primary inductor) creates a flux in the magnetic core, which induces voltage in the other coil (the secondary inductor).

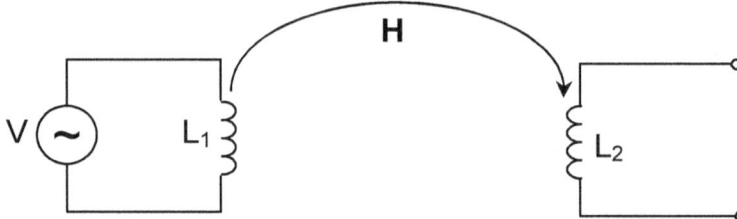

One of the simplest ways to control the flux linkage between the coils is by appropriately increasing or decreasing the number of turns in the inductors relative to one another. In the ideal case, where all the flux produced by the first inductor (L_1 above) is linked to the second inductor (L_2 above), the voltage across L_2 (V_t) is equal to the voltage across L_1 (V) multiplied by the ratio of the number of turns in L_2 (N_2) to the number of turns in L_1 (N_1).

$$V_t = V \frac{N_2}{N_1}$$

Thus, if there are twice as many turns in L_2 as in L_1, twice the flux is linked to L_2 and the voltage is doubled. In reality, however, not all the flux is linked to the second inductor. The use of ferromagnetic cores, especially if a loop of ferromagnetic material is used such that it serves as the core for both inductors, increases the flux linkage and brings the transformer closer to the ideal case. All transformers operate on this simple principle though they range in size and function from those in tiny microphones to those that connect the components of the US power grid.

As a side note, any circuit (which is a wire loop of one form or another) has an inherent inductance that can link the flux from another circuit. Thus, "crosstalk" between circuits can be an undesirable effect that results from the same principles that are used to intentionally design the beneficial transformer device.

Problem: If a step-up transformer has 500 turns on its primary coil and 800 turns on its secondary coil, what will be the output (secondary) voltage be if the primary coil is supplied with 120 V?

Solution:

$$\frac{V_s}{V_p} = \frac{n_s}{n_p}$$

$$V_s = \frac{n_s}{n_p} \times V_p = \frac{800}{500} \times 120V = 192V$$

Skill 15.5 Analyzing an AC circuit, including relationships involving impedance, reactance, and resonance

Alternating current (AC) is a type of electrical current with cyclically varying magnitude and direction. This is differentiated from direct current (DC), which has constant direction. AC is the type of current delivered to businesses and residences.

Though other waveforms are sometimes used, the vast majority of AC current is sinusoidal. Thus we can use wave terminology to help us describe AC current. Since AC current is a function of time, we can express it mathematically as:

$$v(t) = V_{peak} \cdot \sin(\omega t)$$

where V_{peak} = the peak voltage; the maximum value of the voltage

ω = angular frequency; a measure of rotation rate

t = time

A few more terms are useful to help us characterize AC current:

Peak-to-peak value: The difference between the positive and negative peak values. Thus peak-to-peak value is equal to 2 x V_{peak}.

Root mean square value (V_{rms}, I_{rms}): A specific type of average found by the following formulae:

$$V_{rms} = \frac{V_{peak}}{\sqrt{2}} \quad ; \quad I_{rms} = \frac{I_{peak}}{\sqrt{2}} \quad ; \quad I_{rms} = \frac{V_{rms}}{R}$$

V_{rms} is useful because an AC current will deliver the same power as a DC current if its $V_{rms}=V_{DC}$, i.e. average power $P_{av} = V_{rms}I_{rms}$.

Frequency: Describes how often the wave passes through a particular point per unit time. Note that this is physical frequency, f, which is related to the angular frequency ω by:

$$\omega = 2\pi f$$

Impedance: A measure of opposition to an alternating current. It is similar to resistance and also has the unit ohm. However, due to the phased nature of AC, impedance is a complex number, having both real and imaginary components. The resistance is the real part of impedance while the reactance of capacitors and inductors constitute the imaginary part.

Resonant frequency: The frequency at which the impedance between the input and output of the circuit is minimum. At this frequency a phenomenon known as electrical resonance occurs.

Reactance: The impedance contributed by inductors and capacitors in AC circuit. Mathematically, reactance is the imaginary part of impedance. The relationship between impedance (Z), resistance(R), and reactance (X) is given by below.

$$Z = R + Xi$$

1. Remember that i=$\sqrt{-1}$

Problem:

An AC current has V_{rms}=220 V. What is its peak-to-peak value?

Solution:

We simply determine V_{peak} from the definition of V_{rms}:

$$V_{rms} = \frac{V_{peak}}{\sqrt{2}}$$

$$V_{peak} = V_{rms} \times \sqrt{2} = 220V \times \sqrt{2} = 311.12V$$

Therefore,

$$V_{peak-to-peak} = 2 \times V_{peak} = 2 \times 311.12V = 622.24V$$

Skill 15.6 Describing how AC and DC can be converted from one form to another

Conversion of alternating current (AC) to direct current (DC), along with conversion from DC to AC, is important for a number of applications. Devices that require a DC power supply can use an AC source (such as public utilities) when an AC-to-DC converter is used. Similarly, devices that require an AC power supply can use a DC source (such as a battery) when a DC-to-AC converter is used.

AC-to-DC conversion can be obtained through the use of diodes and a filter (a capacitor is often sufficient). A diode only allows current flow in one direction and the filter removes some portion of the AC spectrum (ideally, leaving only the DC level). The use of a transformer to step up or step down the voltage from a particular source (such as the power supply from public utilities) may be required, but this is a separate issue. Below is a schematic for a so-called half-wave rectifier.

Since half the power is lost in this scheme, full-wave rectification is often preferable.

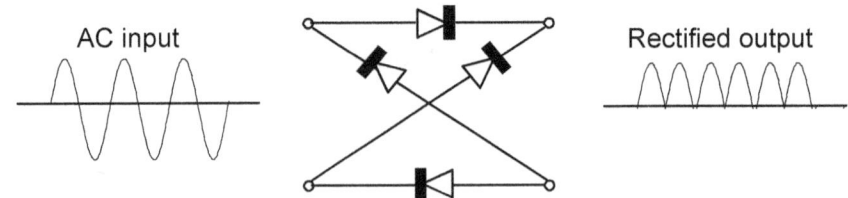

When a capacitor (or other filter) is then added to the full-wave rectifier above, the result is a DC signal with a slight AC ripple.

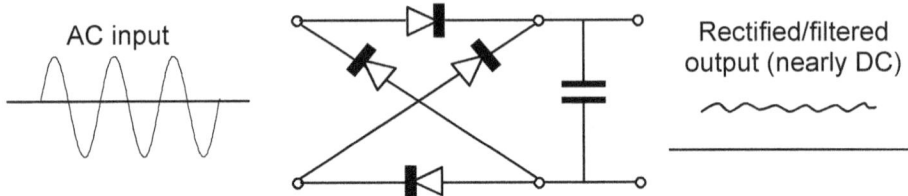

Further filtration can produce a more uniform DC output but this basic schematic shows a simple possibility for reasonably efficient AC-to-DC conversion.

Conversion of DC to AC is a slightly more complicated task. So-called inverters are designed for converting a DC source into an AC signal of some frequency. Again, the use of transformers may be appropriate for variation of the voltage amplitude but only the actual conversion of DC to AC shall be addressed here.

Although inverter design may involve electromechanical or other types of switching to modulate the DC signal the ubiquity of semiconductor devices has made the use of transistors and diodes more prevalent. These inverter circuits can be very complicated, involving a number of semiconductor devices, and can be designed for a wide variety of applications. The electromechanical inverter is easier to comprehend in broad terms, as it can be based on an electromagnet powered by a DC source (such as a battery). A spring-loaded switch is opened when the electromagnet is powered, resulting in a break in the circuit. The electromagnet loses its power as a result of the open circuit, thus closing the switch and powering the electromagnet once more. This new alternating voltage can be tapped and, perhaps after proper filtering, used to provide the AC signal. Such a device, although simple to understand, is very limited in the range of frequencies that can be provided for the AC power. Semiconductor-device-based inverters offer the possibility of much higher frequencies than those offered by electromechanical inverters.

SUBAREA IV   **WAVES, SOUND, AND LIGHT**

COMPETENCY 16.0   UNDERSTAND THE PROPERTIES AND BEHAVIOR OF WAVES

Skill 16.1   Demonstrating knowledge of how waves transfer energy and momentum

Waves are simply disturbances that propagate through space and time. Most waves must propagate through a medium, which may be solid, liquid, or gas. However, some electromagnetic waves do not require a medium and can move through a vacuum. Most waves transfer energy from their source to their destination, typically without actually transferring molecules of the medium through which they travel. The individual particles of the medium are temporarily displaced but eventually return to their equilibrium positions. As the particles oscillate, they transfer energy and momentum to a neighboring particle which, in turn, moves back and forth around its equilibrium position while transferring energy and momentum to another particle and so on.

All around us, there are examples of waves including ocean waves, sound, radio waves, microwaves, seismic waves, sunlight, x-rays, and radioactive gamma rays. Whether these waves actually displace media or simply carry energy, their positions fluctuate as they move through time and space. Often these fluctuations are regular and we can use both diagrams and mathematical equations to understand the pattern a wave follows in space and how quickly it moves.

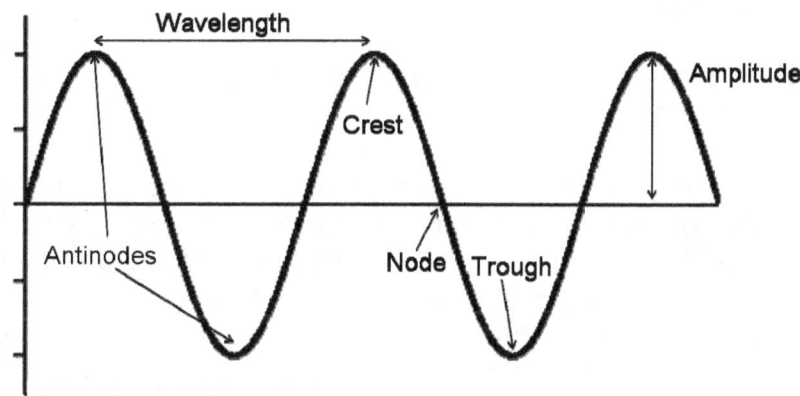

This diagram of a sinusoidal wave shows us displacement caused by the wave as it propagates through a medium. This displacement can be graphed against either time or distance. Note how displacement depends on the distance which the wave has traveled/ how much time has elapsed. So if we chose a particular displacement (let's say the crest), the wave will return to that displacement value (i.e., crest again) after one period (T) or one wavelength (λ).

The general equation for a wave is a partial differential equation which can be simplified to express the behavior of commonly encountered harmonic waveforms (See **Skill 16.4**). For instance, for a standing wave:

$$y(z,t) = A(z,t)\sin(kz - \omega t + \phi)$$

where y=displacement
z=distance
t=time
k=wave number
ω=angular frequency
ϕ=phase
A(z,t)=the amplitude envelope of the wave

The important concept to note is that y is a function of both z and t. This means that the wave's position depends on both time and distance, just as was seen in the diagram above.

Skill 16.2 Comparing types (e.g., longitudinal, transverse) and characteristics (e.g., frequency, period, amplitude, wavelength) of waves

To fully understand waves, it is important to understand many of the terms used to characterize them.

Wave velocity: Two velocities are used to describe waves. The first is phase velocity, which is the rate at which a wave propagates. For instance, if you followed a single crest of a wave, it would appear to move at the phase velocity. The second type of velocity is known as group velocity and is the speed at which variations in the wave's amplitude shape propagate through space. Group velocity is often conceptualized as the velocity at which energy is transmitted by a wave. Phase velocity is denoted v_p and group velocity is denoted v_g.

Crest: The maximum value that a wave assumes; the highest point.

Trough: The lowest value that a wave assumes; the lowest point.

Nodes: The points on a wave with minimal amplitude.

Antinodes: The farthest point from the node on the amplitude axis; both the crests and the troughs are antinodes.

Amplitude: The distance from the wave's highest point (the crest) to the equilibrium point. This is a measure of the maximum disturbance caused by the wave and is typically denoted by A.

Wavelength: The distance between any two sequential troughs or crests denoted λ and representing a complete cycle in the repeated wave pattern.

Period: The time required for a complete wavelength or cycle to pass a given point. The period of a wave is usually denoted T.

Frequency: The number of periods or cycles per unit time (usually a second). The frequency is denoted f and is the inverse of the wave's period (that is, $f=1/T$).

Phase: This is a given position in the cycle of the wave. It is most commonly used in discussing a "being out of phase" or a "phase shift", an offset between waves.

Longitudinal waves: These are created by oscillations in the direction in which the wave travels. Thus, if we imagine a longitudinal waveform moving down a tube, particles will move back and forth parallel to the sides of the tube.

Transverse waves: These waves, on the other hand, oscillate in a direction perpendicular to the direction of wave travel. So let's imagine the same tube, this time with a transverse wave traveling down it. In this case, the particles oscillate up and down or side to side within the tube. Particle displacement in a transverse wave can also be easily visualized in the vibration of a taut string.

Polarization: A property of transverse waves that describes the plane perpendicular to the direction of travel in which the oscillation occurs. Note that longitudinal waves are not polarized because they can oscillate only in one direction, the direction of travel.

Skill 16.3 Applying the wave properties to determine a wave's velocity, wavelength, or frequency

The phase velocity of a wave is related to its wavelength and frequency. Taking light waves, for instance, the speed of light c is equal to the distance traveled divided by time taken. Since the light wave travels the distance of one wavelength λ in the period of the wave T,

$$c = \frac{\lambda}{T}$$

The frequency of a wave, f, is the number of completed periods in one second. In general,

$$f = \frac{1}{T}$$

So the formula for the speed of light can be rewritten as

$$c = \lambda f$$

Thus the phase velocity of a wave is equal to the wavelength times the frequency.

Skill 16.4   Analyzing the transmission and absorption of waves

Electromagnetic waves traveling through a uniform material can be either absorbed or transmitted. Transmission and absorption are complementary phenomena. Reflection, an alternative possibility, may take place if there are non-uniformities in the material and if those non-uniformities are not "matched" (i.e., if they do not both have the same wave impedance).

Maxwell's equations, given here in differential form, provide a time-tested description of electromagnetic waves:

$$\nabla \times \mathbf{E}(\mathbf{r},t) = -\frac{\partial \mathbf{B}(\mathbf{r},t)}{\partial t} \qquad \nabla \times \mathbf{H}(\mathbf{r},t) = \frac{\partial \mathbf{D}(\mathbf{r},t)}{\partial t} + \mathbf{J}(\mathbf{r},t)$$

$$\nabla \cdot \mathbf{D}(\mathbf{r},t) = \rho(\mathbf{r},t) \qquad \nabla \cdot \mathbf{B}(\mathbf{r},t) = 0$$

The two curl equations may be combined, along with either of the divergence equations, to decouple the magnetic and electric fields and to form the wave equation. The wave equation can be expressed in terms of either the magnetic field (**H**) or the electric field (**E**). In this case, for simplicity, homogeneous, isotropic materials are assumed.

$$\nabla^2 \mathbf{H}(\mathbf{r},t) = \mu\varepsilon \frac{\partial^2 \mathbf{H}(\mathbf{r},t)}{\partial t^2} - \nabla \times \mathbf{J}(\mathbf{r},t) \qquad \nabla^2 \mathbf{E}(\mathbf{r},t) = \mu\varepsilon \frac{\partial^2 \mathbf{E}(\mathbf{r},t)}{\partial t^2} + \mu \frac{\partial \mathbf{J}(\mathbf{r},t)}{\partial t}$$

Absorption can usually be modeled as the result of the medium having a finite conductivity. Ideally, in classical electrodynamics, a perfect insulator is expected to transmit all light without any absorption. The presence of mobile charge carriers, such as free or weakly bound electrons, provides a means for absorption, where, for example, electrons acquire energy due to the presence of an electric field (and its corresponding magnetic field). The motion of the electrons may be impeded, resulting in conversion of some of the energy into heat.

In the wave equation, the current density **J** can be expressed as the product of the conductivity σ (a property of the material) and the electric field **E**. The wave equation may be further reduced to the following, where ψ is either **E** or **H**.

$$\nabla^2 \psi(\mathbf{r},t) = \left[\mu\varepsilon\frac{\partial^2}{\partial t^2} + \mu\sigma\frac{\partial}{\partial t}\right]\psi(\mathbf{r},t)$$

This form of the wave equation includes the conductivity σ, and thus accounts for absorption in the material. In the frequency domain, the permittivity ε can be combined with the conductivity to form a generalized complex permittivity, ε_c.

$$\nabla^2 \psi(\mathbf{r}) = \omega^2 \mu \varepsilon_c \psi(\mathbf{r}) \qquad \varepsilon_c = \varepsilon - i\omega\sigma$$

For the simple case of plane waves, the general solution to the above equation is a set of weighted exponentials that represent traveling waves in opposite directions. The general form of the exponential is the following.

$$\psi = e^{i\omega\sqrt{\mu\varepsilon_c}\,r}$$

Since ε is complex, the argument of the exponential is likewise complex and may be expressed as the sum of a real and an imaginary number.

$$\psi = e^{\alpha r} e^{i\beta r}$$

The above expression has two factors: the exp(iβr) term expresses the traveling aspect of the wave; the exp(αr) term expresses the material absorption. The absorption coefficient, α, is generally negative to avoid unphysical wave amplification. In the case where σ = 0, or when the material is non-absorptive, α = 0, and the wave is transmitted without absorption. For materials with a non-zero conductivity, α is finite and there is some amount of absorption, all depending on the material parameters.

Thus, Maxwell's equations can be used to provide a classical explanation of absorption of waves by way of finite material conductivity. Absorption typically leads to an exponential decay of waves, as shown in the case of a simple plane wave. Full transmission occurs in purely (i.e., ideally) non-conductive materials, since no energy is converted to heat.

Skill 16.5 **Analyzing the reflection, refraction, dispersion, diffraction, and polarization of waves**

Wave **refraction** is a change in direction of a wave due to a change in its speed. This most commonly occurs when a wave passes from one material to another, such as a light ray passing from air into water or glass. However, light is only one example of refraction; any type of wave can undergo refraction. Another example would be physical waves passing from water into oil. At the boundary of the two media, the wave velocity is altered, the direction changes, and the wavelength increases or decreases. However, the frequency remains constant.

The **index of refraction**, n, is the amount by which light slows in a given material and is defined by the formula

$$n = \frac{c}{v}$$

where v represents the speed of light through the given material.

Problem: The speed of light in an unknown medium is measured to be $1.24 \times 10^8 \, m/s$. What is the index of refraction of the medium?

Solution:

$$n = \frac{c}{v}$$

$$n = \frac{3.00 \times 10^8}{1.24 \times 10^8} = 2.42$$

Referring to a standard table showing indices of refraction, we would see that this index corresponds to the index of refraction for diamond.

(See **Skill 19.4** for Snell's law of refraction).

Reflection is the change in direction of a wave at an interface between two dissimilar media such that the wave returns into the medium from which it originated. The most common example of this is light waves reflecting from a mirror, but sound and water waves can also be reflected. **The law of reflection states that the angle of incidence is equal to the angle of reflection**.

Diffraction occurs when part of a wave front is obstructed. Diffraction and interference are essentially the same physical process. Diffraction refers to various phenomena associated with wave propagation such as the bending, spreading, and interference of waves emerging from an aperture. It occurs with any type of wave including sound waves, water waves, and electromagnetic waves such as light and radio waves.

Here, we take a close look at important phenomena like single-slit diffraction, double-slit diffraction, diffraction grating, other forms of diffraction and lastly interference.

1. Single-slit diffraction: The simplest example of diffraction is single-slit diffraction in which the slit is narrow and a pattern of semi-circular ripples is formed after the wave passes through the slit.

2. Double-slit diffraction: These patterns are formed by the interference of light diffracting through two narrow slits.

3. Diffraction grating: Diffraction grating is a reflecting or transparent element whose optical properties are periodically modulated. In simple terms, diffraction gratings are fine parallel and equally spaced grooves or rulings on a material surface. When light is incident on a diffraction grating, light is reflected or transmitted in discrete directions, called diffraction orders. Because of their light dispersive properties, gratings are commonly used in monochromators and spectrophotometers. Gratings are usually designated by their groove density, expressed in grooves/millimeter. A fundamental property of gratings is that the angle of deviation of all but one of the diffracted beams depends on the wavelength of the incident light.

4. Other forms of diffraction:
i) Particle diffraction: It is the diffraction of particles such as electrons, which is used as a powerful argument for quantum theory. It is possible to observe the diffraction of particles such as neutrons or electrons and hence we are able to infer the existence of wave particle duality.
ii) Bragg diffraction: This is diffraction from a multiple slits, and is similar to what occurs when waves are scattered from a periodic structure such as atoms in a crystal or rulings on a diffraction grating. Bragg diffraction is used in X-ray crystallography to deduce the structure of a crystal from the angles at which the X-rays are diffracted from it.

5. Interference: Interference is described as the superposition of two or more waves resulting in a new wave pattern. Interference is involved in Thomas Young's double slit experiment where two beams of light which are coherent with each other interfere to produce an interference pattern. Light from any source can be used to obtain interference patterns. For example, Newton's rings can be produced with sun light. However, in general, white light is less suited for producing clear interference patterns as it is a mix of a full spectrum of colors. Sodium light is close to monochromatic and is thus more suitable for producing interference patterns. The most suitable is laser light as it is almost perfectly monochromatic.

Dispersion is the separation of a wave into its constituent wavelengths due to interaction with a material occurring in a wavelength-dependent manner (as in thin-film interference for instance). Dispersive prisms separate white light into these constituent colors by relying on the differences in refractive index that result from the varying frequencies of the light. Prisms rely on the fact that light changes speed as it moves from one medium to another. This then causes the light to be bent and/or reflected. The degree to which bending/reflection occurs is a function of the light's angle of incidence and the refractive indices of the media.

Polarization is a property of transverse waves that describes the plane perpendicular to the direction of travel in which the oscillation occurs. In unpolarized light, the transverse oscillation occurs in all planes perpendicular to the direction of travel. Polarized light (created, for instance, by using polarizing filters that absorb light oscillating in other planes) oscillates in only a selected plane. An everyday example of polarization is found in polarized sunglasses which reduce glare.

Skill 16.6 Applying the superposition principle to determine characteristics of a resultant wave

According to the **principle of linear superposition**, when two or more waves exist in the same place, the resultant wave is the sum of all the waves, i.e. the amplitude of the resulting wave at a point in space is the sum of the amplitudes of each of the component waves at that point. Interference is usually observed in coherent waves, well-correlated waves that have very similar frequencies or even come from the same source.

Superposition of waves may result in either constructive or destructive interference. **Constructive interference** occurs when the crests of the two waves meet at the same point in time. Conversely, **destructive interference** occurs when the crest of one wave and the trough of the other meet at the same point in time. It follows, then, that constructive interference increases amplitude and destructive interference decreased it. We can also consider interference in terms of wave phase; waves that are out of phase with one another will interfere destructively while waves that are in phase with one another will interfere constructively. In the case of two simple sine waves with identical amplitudes, for instance, amplitude will double if the waves are exactly in phase and drop to zero if the waves are exactly 180° out of phase.

Additionally, **interference can create a standing wave**, a wave in which certain points always have amplitude of zero. Thus, the wave remains in a constant position. Standing waves typically results when two waves of the same frequency traveling in opposite directions through a single medium are superposed. View an animation of how interference can create a standing wave at the following URL:

http://www.glenbrook.k12.il.us/GBSSCI/PHYS/mmedia/waves/swf.html

All wavelengths in the EM spectrum can experience interference but it is easy to comprehend instances of interference in the spectrum of visible light. One classic example of this is Thomas Young's double-slit experiment. In this experiment a beam of light is shone through a paper with two slits and a striated wave pattern results on the screen. The light and dark bands correspond to the areas in which the light from the two slits has constructively (bright band) and destructively (dark band) interfered.

Similarly, we may be familiar with examples of interference in sound waves. When two sounds waves with slightly different frequencies interfere with each other, beat results. We hear a beat as a periodic variation in volume with a rate that depends on the difference between the two frequencies. You may have observed this phenomenon when listening to two instruments being tuned to match; beating will be heard as the two instruments approach the same note and disappear when they are perfectly in tune.

Skill 16.7   Solving problems involving wave phenomena

Because mechanical, sound, and light waves surround us everyday, there are a variety of situations available for us to analyze. Though problems involving optics are more complicated (See **Skills 19.1-19.3**), we can solve simple problems involving waves by applying the definitions and equations from above. Below are several examples.

Problem:
A wave travels along a string. If it takes 0.30 seconds to move from its lowest to highest point, what is the frequency of the wave? If the wavelength is 1.8 m, what is the speed of the wave?

Solution:
Since 0.30 seconds are required for the wave to move from its lowest to highest point, the period must be 0.60 seconds. We can find frequency from period:

$$f = \frac{1}{T} = \frac{1}{0.6\sec} = 1.67\sec^{-1} = 1.67 Hz$$

We already know the frequency and wavelength of the wave, so we can easily find the speed:

$$v = f \times \lambda = 1.67 Hz \times 1.8 m = 3.0 \frac{m}{\sec}$$

Problem:
Waves are generated in two identical tanks of water. The amplitude of the wave in Tank 1 is 0.4 cm; in Tank 2 the amplitude is 0.8 cm. How much more energy is transported by the wave in Tank 2 than in Tank 1?

Solution:
Remember that the energy of a wave is proportional to the square of its amplitude. Thus, doubling the amplitude quadruples the energy, and the wave in Tank 2 transports four times as much energy as the one in Tank 1.

Problem:
A woman hiking in a canyon yells out "hello" and 2.4 seconds later hears her echo (the reflection of the sound wave off the canyon wall). If the speed of the sound wave is 345 m/sec, how far away is the canyon wall?

Solution:
If it takes 2.4 seconds for the echo to reach the woman, it must take half that time for the sound to reach canyon wall, that is, 1.2 seconds. Thus, we can find the distance to the wall:

$$d = v \times t = 345 \frac{m}{\sec} \times 1.2 \sec = 414 m$$

Problem:
Two tuning forks are played simultaneously. If one fork has a frequency of 450 Hz and the other has a frequency of 446 Hz, how many beats will be heard after 10 seconds?

Solution
The beat frequency will be 450-446 Hz= 4 Hz

$$Number\ of\ beats = 4 Hz \times 10 \sec = 40\ beats$$

Problem: The interference maxima (location of bright spots created by constructive interference) for double-slit interference are given by

$$\frac{n\lambda}{d} = \frac{x}{D} = \sin\theta \quad n=1,2,3...$$

where λ is the wavelength of the light, d is the distance between the two slits, D is the distance between the slits and the screen on which the pattern is observed and x is the location of the nth maximum. If the two slits are 0.1mm apart, the screen is 5m away from the slits, and the first maximum beyond the center one is 2.0 cm from the center of the screen, what is the wavelength of the light?

PHYSICS

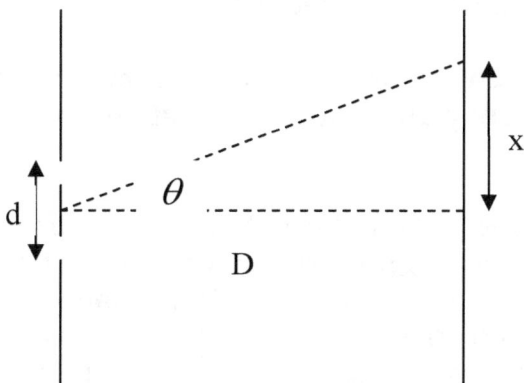

Solution: $\lambda = xd/(Dn) = 0.02 \times 0.0001/ (5 \times 1) = 400$ nanometers

TEACHER CERTIFICATION STUDY GUIDE

COMPETENCY 17.0 UNDERSTAND AND APPLY KNOWLEDGE OF THE CHARACTERISTICS, PRODUCTION, AND TRANSMISSION OF SOUND WAVES

Skill 17.1 Analyzing the physical nature of sound waves, including intensity and frequency and how they relate to loudness and pitch, respectively

Sound waves are mechanical waves that can travel through various kinds of media, solid, liquid and gas. They are **longitudinal waves** transmitted as variations in the pressure of the surrounding medium. These variations in pressure form waves that are termed sound waves or acoustic waves. When the particles of the medium are drawn close together it is called **compression**. When the particles of the medium are spread apart it is known as **rarefaction**.

Sound waves have different characteristics in different materials (see **Skill 17.2**). Boundaries between different media can result in partial or total reflection of sound waves. Thus, the phenomenon of echo takes place when a sound wave strikes a material of differing characteristics than the surrounding medium. The reflection of a voice off the walls of a room is a particular example.

The frequency of the harmonic acoustic wave, ω, determines the **pitch** of the sound in the same manner that the frequency determines the color of an electromagnetic harmonic wave. **A high pitch sound corresponds to a high frequency sound wave** and a low pitch sound corresponds to a low frequency sound wave. Similarly, the amplitude of the pressure determines the **loudness** or just as the amplitude of the electric (or magnetic) field **E** determines the brightness or intensity of a color. **The intensity of a sound wave is proportional to the square of its amplitude.**

The decibel scale is used to measure sound intensity. It originated in a unit known as the bel which is defined as the reduction in audio level over 1 mile of a telephone cable. Since the bel describes such a large variation in sound, it became more common to use the decibel, which is equal to 0.1 bel. A decibel value is related to the intensity of a sound by the following equation:

$$X_{dB} = 10\log_{10}\left(\frac{X}{X_0}\right)$$

Where X_{dB} is the value of the sound in decibels
       X is the intensity of the sound
       X_0 is a reference value with the same units as X. X_0 is commonly taken to be the threshold of hearing at $10^{-12}\,W/m^2$.

It is important to note the logarithmic nature of the decibel scale and what this means for the relative intensity of sounds. The perception of the intensity of sound increases logarithmically, not linearly. Thus, an increase of 10 dB corresponds to an increase by one order of magnitude. For example, a sound that is 20 dB is not twice as loud as sound that is 10 dB; rather, it is 10 times as loud. A sound that is 30 dB will the 100 times as loud as the 10 dB sound.

Finally, let's equate the decibel scale with some familiar noises. Below are the decibel values of some common sounds.

Whispering voice: 20 dB
Quiet office: 60 dB
Traffic: 70 dB
Cheering football stadium: 110 dB
Jet engine (100 feet away): 150 dB
Space shuttle liftoff (100 feet away): 190 dB

Skill 17.2 Analyzing factors that affect the speed of sound in different media

Mechanical waves rely on the local oscillation of each atom in a medium, but the material itself does not move; only the energy is transferred from atom to atom. Therefore the material through which the mechanical wave is traveling greatly affects the wave's propagation and speed. In particular, a material's elastic constant and density affect the speed at which a wave travels through it. Both of these properties of a medium can predict the extent to which the atoms will vibrate and pass along the energy of the wave. The general relationship between these properties and the speed of a wave in a solid is given by the following equation:

$$V = \sqrt{\frac{C_{ij}}{\rho}}$$

where V is the speed of sound, C_{ij} is the elastic constant or bulk modulus, and ρ is the material density. It is worth noting that the elastic constant differs depending on direction in anisotropic materials and the ij subscripts indicate that this directionality must be taken into account.

The speed of sound also varies with temperature since material characteristics are often dependent on temperature. At $0^\circ C$, the speed of sound is 331m/s in air, 1402m/s in water and 6420m/s in Aluminum. Even though water is much denser than air, it is a lot more incompressible than air and its bulk modulus is larger than that of air by a larger factor. Therefore sound travels much faster in water than in air.

In the case of a stretched string, the velocity of a transverse wave passing through it is given by

$$v = \sqrt{\frac{\tau}{\mu}}$$

where τ is the tension in the string and μ is its linear density.

Skill 17.3   Analyzing characteristics of standing waves in musical instruments (e.g., winds, strings, percussion)

A **standing wave** typically results from the interference between two waves of the same frequency traveling in opposite directions. The result is a stationary vibration pattern. One of the key characteristics of standing waves is that there are points in the medium where no movement occurs. The points are called **nodes** and the points where motion is maximal are called **antinodes**. Thus the amplitude of a standing wave varies with location. This property allows for the analysis of various typical standing waves.

Strings

Imagine a string of length L tied tightly at its two ends. We can generate a standing wave by plucking the string. The waves traveling along the string are reflected at the fixed end points and interfere with each other to produce standing waves. There will always be two nodes at the ends where the string is tied. Depending on the frequency of the wave that is generated, there may also be other nodes along the length of the string. In the diagrams below, examples are given of strings with 0, 1, or 2 additional nodes.

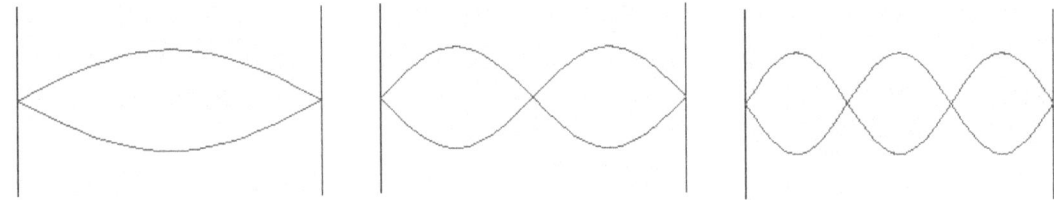

Percussion

On a percussion instrument such as a drum, standing waves occur on the two-dimensional surface in many different configurations. The nodes are circles and straight lines rather than points as on a string. Some possible node patterns are shown below.

Winds

Just as on a string, standing waves can propagate in gaseous or liquid medium inside a tube. In these cases we will also observe harmonic vibrations, but their nature will depend on whether the ends of the tube are closed or open. Specifically, an antinode will be observed at an open end and a node will appear at a closed end. As in strings, there may be additional nodes in between.

Skill 17.4 Analyzing situations involving resonance, harmonics, and overtones

Due to the presence of nodes in a standing wave pattern, a stretched string or a tube with ends closed vibrates at certain characteristic frequencies that correspond to different node patterns. These frequencies are known as **harmonics** or characteristic **resonance frequencies** of the instrument. The instrument will absorb much greater energy when excited at one of the resonance frequencies. The lowest of these frequencies is known as the **fundamental** while the higher ones are known as **overtones**. Harmonic frequencies for vibrating strings and air tubes are calculated below.

Vibrating string

Since we know the length of the string (L) in each case, we can calculate the wavelength (λ) and frequency (f) for any harmonic using the following formula, where n=the harmonic order (n=1,2,3...) and v is the phase velocity of the wave.

$$\lambda_n = \frac{2L}{n} \qquad f_n = \frac{v}{\lambda_n} = n\frac{v}{2L}$$

Waves in a tube
For a tube with two closed ends or two open ends (note that this is the same as for the string described above):

$$f_n = \frac{nv}{2L}$$

where n = 1,2,3,4...

When only one end of a tube is closed, that end become a node and wave exhibits odd harmonics. For a tube with one close end and one open end:

$$f_n = \frac{nv}{4L}$$

where n = 1,3,5,7...

Wavelengths can be found by applying the formula f=v/λ.

Problem:
A string on a violin is 75 cm long. The speed of waves produced when the string is played is known to be 345 m/s. What is the fundamental frequency of the string?

Solution:
Remember that fundamental frequency is simply the first harmonic of the string. The wavelength of the first harmonic must be twice the length of the string:

$$\lambda = 2 \times l = 2 \times 0.75m = 1.5m$$

Then find the fundamental frequency:

$$f = \frac{v}{\lambda} = \frac{345\frac{m}{s}}{1.5m} = 230s^{-1} = 230Hz$$

Skill 17.5   Analyzing and solving problems involving the Doppler effect

The Doppler effect is the name given to the perceived change in frequency that occurs when the observer or source of a wave is moving. Specifically, the perceived frequency increases when a wave source and observer move toward each other and decreases when a source and observer move away from each other. Thus, the source and/or observer velocity must be factored in to the calculation of the perceived frequency. The mathematical statement of this effect is:

$$f' = f_0 \left(\frac{v \pm v_o}{v \pm v_s} \right)$$

where f'= observed frequency
f_0= emitted frequency
v= the speed of the waves in the medium
v_s= the velocity of the source (positive in the direction away from the observer)
v_o= the velocity of the observer (positive in the direction towards the source)

Note that any motion that changes the perceived frequency of a wave will cause the Doppler effect to occur. Thus, the wave source, the observer position, or the medium through which the wave travels could possess a velocity that would alter the observed frequency of a wave. You may view animations of stationary and moving wave sources at the following URL:

http://www.kettering.edu/~drussell/Demos/doppler/doppler.html

So, let's consider two examples involving sirens and analyze what happens when either the source or the observer moves. First, imagine a person standing on the side of a road and a police car driving by with its siren blaring. As the car approaches, the velocity of the car will mean sound waves will "hit" the observer as the car comes closer and so the pitch of the sound will be high. As it passes, the pitch will slide down and continue to lower as the car moves away from the observer. This is because the sound waves will "spread out" as the source recedes. Now consider a stationary siren on the top of fire station and a person driving by that station. The same Doppler effect will be observed: the person would hear a high frequency sound as he approached the siren and this frequency would lower as he passed and continued to drive away from the fire station.

Problem: A sound source moves towards an observer at 8m/s while the observer moves towards the source at 9m/s. If the observer detects a sound of frequency 83kHz, what is the frequency of the sound emitted? The speed of sound in the medium is 343 m/s.

Solution: Using the Doppler equation, the emitted frequency f is given by

$$83 = f\frac{343+9}{343-8}; \Rightarrow f = 79kHz$$

COMPETENCY 18.0 UNDERSTAND THE PRODUCTION AND CHARACTERISTICS OF ELECTROMAGNETIC WAVES

Skill 18.1 Analyzing the properties (e.g., energy, frequency, wavelength) of components (e.g., colors of visible light, ultraviolet radiation) of the electromagnetic spectrum

The electromagnetic spectrum is measured using frequency (f) in hertz or wavelength (λ) in meters. The frequency times the wavelength of every electromagnetic wave equals the speed of light (3.0×10^8 meters/second).

Roughly, the range of wavelengths of the electromagnetic spectrum is:

	f	λ
Radio waves	$10^{5} - 10^{-1}$ hertz	$10^{3} - 10^{9}$ meters
Microwaves	$3 \times 10^{9} - 3 \times 10^{11}$ hertz	$10^{-3} - 10^{-1}$ meters
Infrared radiation	$3 \times 10^{11} - 4 \times 10^{14}$ hertz	$7 \times 10^{-7} - 10^{-3}$ meters
Visible light	$4 \times 10^{14} - 7.5 \times 10^{14}$ hertz	$4 \times 10^{-7} - 7 \times 10^{-7}$ meters
Ultraviolet radiation	$7.5 \times 10^{14} - 3 \times 10^{16}$ hertz	$10^{-8} - 4 \times 10^{-7}$ meters
X-Rays	$3 \times 10^{16} - 3 \times 10^{19}$ hertz	$10^{-11} - 10^{-8}$ meters
Gamma Rays	$> 3 \times 10^{19}$ hertz	$< 10^{-11}$ meters

Radio waves are used for transmitting data. Common examples are television, cell phones, and wireless computer networks. Microwaves are used to heat food and deliver Wi-Fi service. Infrared waves are utilized in night vision goggles. Visible light we are all familiar with as the human eye is most sensitive to this wavelength range. UV light causes sunburns and would be even more harmful if most of it were not captured in the Earth's ozone layer. X-rays aid us in the medical field and gamma rays are most useful in the field of astronomy.

Skill 18.2 Analyzing color when light is transmitted, absorbed, and reflected

A number of different factors affect the transmission, absorption and reflection of light. While these may be examined from a mathematical perspective, color is a psychological phenomenon that is dependent as much on subjective perception as on objective factors such as the wavelength and intensity of the light. Nevertheless, there is a sufficient relationship between color and the objective properties of light, as determined by such factors as wavelength, so as to make an objective analysis appropriate to a significant extent.

The color of a material is determined largely by its reflection, absorption and transmission characteristics, as well as the spectrum of the illuminant. For simplicity of explanation, it may be assumed for the moment that the illuminant has a spectrum that at least corresponds nearly to "white light." If an object is illuminated uniformly, then its color (or colors, if the object includes a number of different materials) is determined by the light that is reflected and the light that is transmitted.

As light travels through a medium, some portion may be absorbed, depending on the material properties and, concomitantly, on the wavelength of the light. That portion of the light (ray) that is not absorbed is transmitted and received by the eye, and brain interpretation of the light as color follows. Some wavelengths may be absorbed more readily by a particular material than others, in which case the material may have a characteristic color. Stained glass is one example of this phenomenon, where additives in different kinds of glass more readily absorb certain wavelengths, thus leaving others to transmit and give the glass its characteristic color.

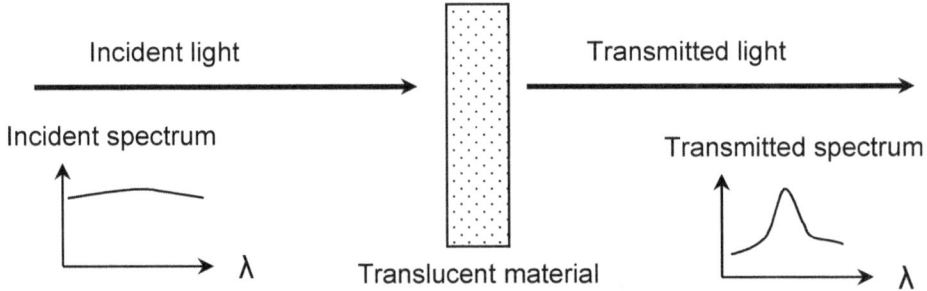

Of course, full absorption of incident light results in a lack of color in the object so long as the illuminant remains "behind" it. Full transmission makes the object appear clear (although other properties of a material could potentially create distortions, preventing the object from being "invisible").

Reflected light is the other component of the physical aspect of color, and absorption can play a similar role to that played with regard to transmission. The light that is neither absorbed nor transmitted is reflected. Reflected light may have the same spectrum as the illuminant, as is often the case for so-called specular reflection. In this situation, for certain viewing angles, the object shows a highlight that is usually the same color as the illuminant (although for some materials the specular highlight has a different spectrum, as with the reddish highlights of copper). Thus, for example, "sun glare" on the surface of water shows the approximately white light of the sun rather than the bluish color of the water. Specular reflection tends to occur more readily on smooth surfaces (e.g., glossy surfaces); diffuse reflection tends to occur more readily on rough surfaces (e.g., matte surfaces).

When only a portion of the spectrum of incident light is reflected, the object can have a characteristic color. This phenomenon is similar to that of transmission.

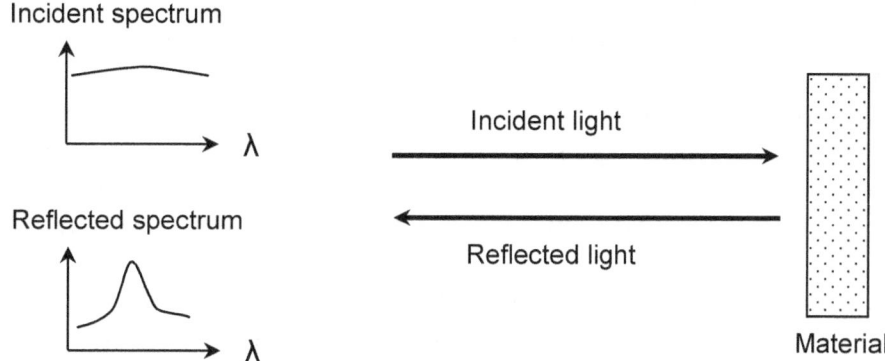

The spectrum of the reflected light is determined by the types of reflection involved (specular versus diffuse), as well as the level of absorption or transmission. These factors, and thus the color, may also vary with viewing angle. As with prisms, some wavelengths may be transmitted or reflected at different angles from those of other wavelengths. Thus, a compact disk can reveal a number of different colors, as well as significant specular reflection, depending on the angle from which it is viewed.

Skill 18.3   Analyzing variations in energy, frequency, and amplitude in terms of the vibrations of the sources that produce them (e.g., molecules, electrons, nuclear particles)

Vibrations, as opposed to unidirectional motion, require an unbalancing force followed by a restoring force. A common macroscopic example is a spring harmonic oscillator with a mass suspended from a stationary surface by way of a spring. If the mass is pulled down further (an unbalancing force), the tension of the spring increases as the mass is displaced until an equilibrium is reached. Once released, the mass will return to its original location; at this point, however, the energy used to displace the mass has been converted from potential energy to kinetic energy. Although the forces are balanced at the point of zero initial displacement, the kinetic energy of the mass causes it to pass the initial location, thus converting kinetic energy to potential energy once more as a restoring force increases in an effort to return the mass to its initial location. This process repeats, resulting in oscillations.

A similar process can take place at the microscopic level with atoms or molecules in a material. As the frequency and amplitude of the constituent elements of the material increase, the result is a corresponding increase in thermal energy. This increase in energy can be registered as an increase in temperature.

When charged particles or molecules acquire vibrational energy, the energy can be radiated as electromagnetic waves (or photons). Electrons are a simple example. An ensemble of electrons in a metal rod, for instance, can vibrate as the result of an applied alternating electric field. This causes an alternating current, which, according to Maxwell's equations, leads to a propagating electromagnetic field. The frequency of oscillation determines the frequency of the wave. If the charged particle is in a bulk material, the energy from oscillation can either be emitted as electromagnetic radiation, as mentioned, or as a phonon, a vibrational mode that is treated like a particle. This is a more refined way of addressing thermal energy, but is similar in terms of the basic concepts.

From a quantum mechanical perspective, oscillations (such as those of electrons) can be dealt with by solving the Schrödinger wave equation using the quantum mechanical Hamiltonian. This Hamiltonian is similar in form to the expression for the mechanical energy of a classical oscillator and includes a term for kinetic energy and a term for potential energy. Instead of producing a continuous spectrum of possible energies, however, the quantum mechanical approach yields a discrete set of energy levels for the oscillator. Changes between energy levels require the emission (or absorption) of a specific discrete amount of energy. Such changes can occur through the emission or absorption of a photon or phonon.

Skill 18.4 Analyzing practical applications of the components of the electromagnetic spectrum (e.g., infrared detectors, solar heating, x-ray machines, AM and FM radio signals, holographic images)

The characteristics of electromagnetic radiation vary depending on the frequency of the radiation and the properties of the material through which it is propagating. Virtually innumerable applications have been devised for various frequency bands, leading to the development of a number of interesting and practical devices and inventions.

A common application of the microwave portion of the electromagnetic spectrum is the microwave oven. Since water molecules have a local charge imbalance, they each manifest a dipole moment. Microwave radiation operates at a wavelength such that incidence on water molecules imparts oscillating motion because of the dipole moments. This motion causes the temperature of the water to rise, thus cooking the food.

Another useful application of the electromagnetic spectrum is remote communications. A number of different frequency bands can be exploited, each with its own characteristics. Since oscillation of charges produces electromagnetic waves that propagate away from the source, antennas that are supplied with an electrical signal can be used to send or receive information through modulation of the waves. Two common modulation schemes are amplitude modulation (AM), for which the amplitude of a sine wave at a given frequency is varied, and frequency modulation (FM), for which the frequency of a sine wave (with a specific central frequency) is modulated. In both cases, these modulation schemes use a band of frequencies (bandwidth) in accordance with Fourier theory. Thus, a broadcast radio station is actually allocated a range of frequencies around a single central frequency (50 kHz around 100.5 MHz, for example). A virtually unlimited range of frequencies can be used for communication although some are more suited to certain situations than others. For example, line-of-sight communication is amenable to the use of a visible (or non-visible) laser as well as microwaves which are commonly used for cellular phones.

In cases where visible light is dim or absent, infrared detectors or video equipment can be used to produce images of scenes that appear dark to the unaided eye. Depending on their temperature, objects radiate infrared waves which can be used to produce false-color images of an otherwise dark environment.

Another method of imaging is the use of X-rays. These high-energy (i.e., high frequency) waves are able to penetrate certain materials, such as flesh, but are unable to penetrate other materials, such as bone or metal. Thus, X-rays may be used to visualize the skeletal structure of a human or animal without an invasive procedure or they may be used to detect hidden metallic items, such as various forms of contraband.

An advanced form of imaging is holography, which allows recording of a three-dimensional image, rather than just a two-dimensional image. Holography uses the coherent light produced by lasers to imprint an image on a photographic plate; in addition to amplitude information, which is all that is recorded in standard photography, holography also includes phase information, thus allowing capture of a three-dimensional image.

The broad range of frequencies radiated by the Sun can be collected in a number of ways to beneficially use or convert the incident energy. So-called solar panels can be used to convert electromagnetic radiation into electricity for powering devices or for storage in batteries. Also, the heat produced as a result of absorption of solar radiation can be used, for example, to heat a house. An appropriate system to collect the radiation, convert it to heat and distribute it throughout a house or other structure is one particular use of this phenomenon.

Skill 18.5 Describing how the constancy of the speed of light in a vacuum influences relationships among space, time, and energy

One of the postulates of Einstein's theory of special relativity is that the speed of light c is invariant with respect to inertial frames of reference. The theory thus requires that, regardless of the relative motion of the source, the observer will always measure the same value for the speed of light. This has numerous implications for the relationships among space, time and energy. These concepts, which are generally independent in Newtonian (classical) mechanics, are found to be related in relativistic mechanics through the Lorentz transformation equations.

Space-time
The implications of the constancy of c can be seen through a comparison of the various parameters as viewed from two different inertial frames of reference, S and S'. Let frame S be stationary and frame S' have a velocity v along the x-axis. The Lorentz transformations require that an event at time t and position (x, y, z) in frame S must occur at time t' and position (x', y', z') in frame S'.

$$t' = \gamma \left(t - \frac{vx}{c^2} \right)$$
$$x' = \gamma (x - vt)$$
$$y' = y$$
$$z' = z$$

Here, the Lorentz factor γ is defined as follows.

$$\gamma = \frac{1}{\sqrt{1 - \frac{v^2}{c^2}}}$$

The Lorentz transformation equations shows that, contrary to classical mechanics, where t' = t, time and space are codependent in the context of special relativity. As a result, rather than referring to time and space as separate aspects of the universe, physicists will often speak of space-time, a singular characteristic which is manifested in both spatial and temporal dimensions. The following expression, which is invariant with respect to the Lorentz transformation, has the characteristics of a "length" of a vector in space-time.

$$x^2 + y^2 + z^2 - c^2 t^2$$

The interconnectedness of space and time (i.e., space-time), results in counterintuitive phenomena such as length contraction and time dilation. In daily life, since the objects encountered are traveling at speeds much less than c, such effects are not seen. For these circumstances, the Lorentz transformations reduce approximately to the classical transformations, where all dimensions in frame S are the same as those in frame S', with the exception of x' = x – vt.

Energy

The constancy of the speed of light also has implications with regard to energy. The relativistic energy of an object is given below.

$$E^2 = (pc)^2 + (mc^2)^2$$

Here, p is the momentum of the object. This equation shows that energy and mass are not distinct parameters, but are related. Furthermore, since energy and momentum are dependent upon the inertial frame of reference, mass is likewise dependent upon the inertial frame of reference. The mass of an object, m, varies with velocity according to the Lorentz factor, as given below. The rest mass m_0 is used here.

$$m = \gamma m_0$$

Thus, it can be seen that the constancy of the speed of light leads to unique, and sometimes difficult to comprehend, relationships between mass, energy, space and time, such as are not seen in the simpler case of classical mechanics.

Skill 18.6 Analyzing and solving problems involving the Doppler effect

The Doppler effect is observed with all types of electromagnetic radiation as in the case of sound (see **Skill 17.5**). The frequency or wavelength shift in the case of electromagnetic radiation is observed as a change in color. The equations, however, are only identical to that of sound when the relative speed of source and observer is much less than that of light.

The Doppler effect for light has been exploited by astronomers to measure the speed at which stars and galaxies are approaching or moving away. In astronomy, Doppler effect is measured in terms of wavelength rather than frequency. An approximate equation in cases where the relative velocity of source and observer is very much smaller than light speed c is

$$u = \frac{\Delta \lambda}{\lambda}$$

where u is the relative velocity of the source and observer, $\Delta \lambda$ is the change in wavelength of the observed light and λ is the actual wavelength of the light. If the wavelength decreases this is known as a **blue shift** since the frequency increases and the highest visible frequencies are in the blue range. An increase in wavelength is known as a **red shift**.

COMPETENCY 19.0   UNDERSTAND AND APPLY THE CHARACTERISTICS OF REFLECTIVE AND REFRACTIVE DEVICES

Skill 19.1   Comparing types and characteristics of lenses and mirrors

Lenses

A lens is a device that causes electromagnetic radiation to converge or diverge. The most familiar lenses are made of glass or plastic and designed to concentrate or disperse visible light. Two of the most important parameters for a lens are its thickness and its focal length. Focal length is a measure of how strongly light is concentrated or dispersed by a lens. For a convex or converging lens, the focal length is the distance at which a beam of light will be focused to a single spot. Conversely, for a concave or diverging lens, the focal length is the distance to the point from which a beam appears to be diverging.

The images produced by lenses can be either virtual or real. A virtual image is one that is created by rays of light that appear to diverge from a certain point. Virtual images cannot be seen on a screen because the light rays do not actually meet at the point where the image is located. If an image and object appear on the same side of a converging lens, that image is defined as virtual. For virtual images, the image location will be negative and the magnification positive. Real images, on the other hand, are formed by light rays actually passing through the image. Thus, real images are visible on a screen. Real images created by a converging lens are inverted and have a positive image location and negative magnification.

Plane mirrors

Plane mirrors form virtual images. In other words, the image is formed behind the mirror where light does not actually reach. The image size is equal to the object size and object distance is equal to the image distance; i.e. the image is the same distance behind the mirror as the object is in front of the mirror. Another characteristic of plane mirrors is left-right reversal.

Example: Suppose you are standing in front of a mirror with your right hand raised. The image in the mirror will be raising its left hand.

Problem: If a cat creeps toward a mirror at a rate of 0.20 m/s, at what speed will the cat and the cat's image approach each other?

Solution: In one second, the cat will be 0.20 meters closer to the mirror. At the same time, the cat's image will be 0.20 meters closer to the cat. Therefore, the cat and its image are approaching each other at the speed of 0.40 m/s.

Problem: If an object that is two feet tall is placed in front of a plane mirror, how tall will the image of the object be?

Solution: The image of the object will have the same dimensions as the actual object, in this case, a height of two feet. This is because the magnification of an image in a plane mirror is 1.

Curved mirrors

Curved mirrors are usually sections of spheres. In a concave mirror the inside of the spherical surface is silvered while in a convex mirror it is the outside of the spherical surface that is silvered.

Terminology associated with spherical mirrors:

Principal axis: The line joining the center of the sphere (of which we imagine the mirror is a section) to the center of the reflecting surface.

Center of curvature: The center of the sphere of which the mirror is a section.

Vertex: The point on the mirror where the principal axis meets the mirror or the geometric center of the mirror.

Focal point: The point at which light rays traveling parallel to the principal axis will meet after reflection in a concave mirror. For a convex mirror, it is the point from which light rays traveling parallel to the principal axis will appear to diverge from after reflection. The focal point is midway between the center of curvature and the vertex.

Focal length: The distance between the focal point and the vertex.

Radius of curvature: The distance between the center of curvature and the vertex, i.e. the radius of the sphere of which the mirror is a section. The radius of curvature is twice the focal length.

Image characteristics for **concave mirrors**:
i. If the object is located **beyond the center of curvature**, the image will be **real, inverted, smaller** and located between the focal point and center of curvature.
ii. If the object is located **at the center of curvature**, the image will be **real, inverted, of the same height** and also located at the center of curvature.
iii. If the object is located **between the center of curvature and focal point**, the image will be **real, inverted, larger** and located beyond the center of curvature.
iv. If the object is located **at the focal point no image is formed**.
v. If the object is located **between the focal point and vertex**, the image will be **virtual, upright, larger** and located on the opposite side of the mirror.
vi. If the object is located **at infinity** (very far away), the image is **real, inverted, smaller** and located at the focal point.

For **convex mirrors**, the image is always **virtual, upright, reduced in size** and formed on the opposite side of the mirror.

Skill 19.2 Using a ray diagram to locate the focal point or point of image formation of a lens or mirror

Ray diagrams are a convenient way to visualize the propagation of waves and to perform reasonably accurate calculations of the effect of mirrors and lenses on light.

For mirrors, the focal point is either real (for concave mirrors) or virtual (for convex mirrors). In either case, the focal point is found by looking at the behavior of two parallel rays incident upon the mirror.

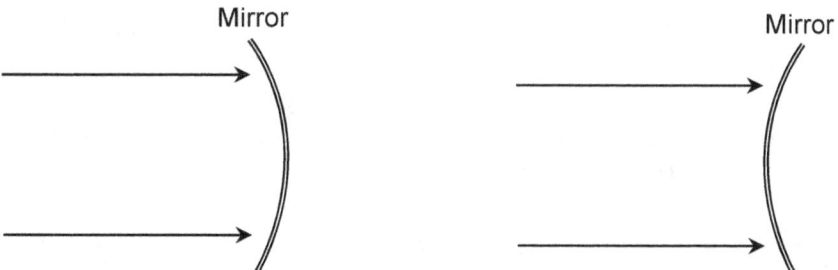

For each incident ray, the angle of reflection θ_r is equal to the angle of incidence θ_r, where the angle is measured from the normal to the mirror surface at the point of incidence.

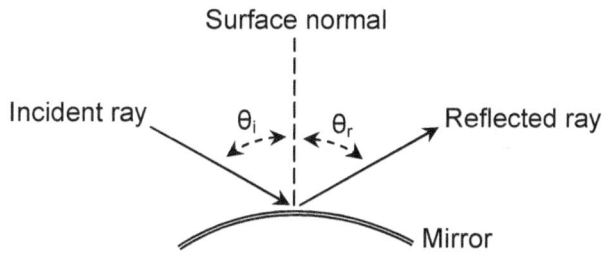

When this law is applied to the rays for either mirror, the focal points (f) are revealed as the (real or virtual) intersection of the rays. The virtual focus point is the intersection for the reflected rays when they are extended beyond the surface of the mirror.

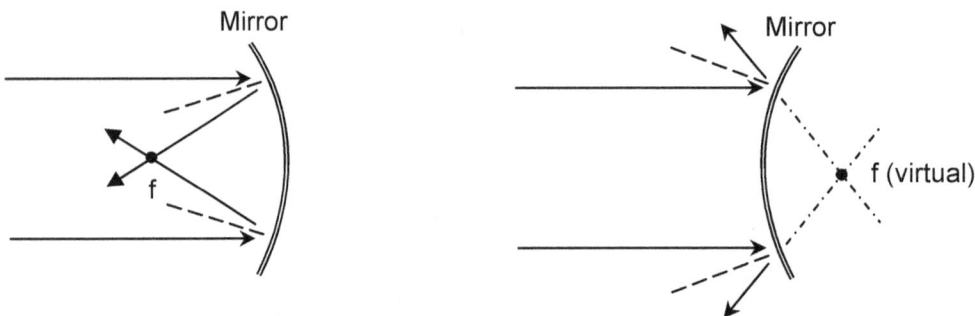

The focal point of a lens is found in a similar manner, with the exception that instead of using the law of reflection, refraction by way of Snell's law must be applied. Since real lenses have a finite thickness, Snell's law must be used for the ray both as it enters the lens material and as it exits the lens material.

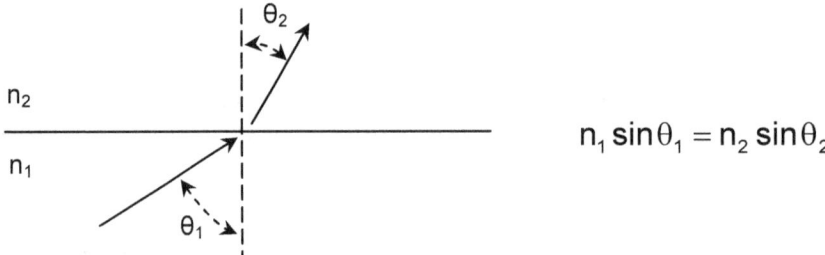

$$n_1 \sin\theta_1 = n_2 \sin\theta_2$$

As with reflection, refraction requires consideration of the normal to the surface. Also, the refractive indices must be used. It is assumed here that the refractive index of the outer medium is n_1 and the refractive index of the lens is n_2, and that $n_1 < n_2$. The intersection of parallel incident rays determines the focal point, which may again be either real or virtual. Real focal points occur on the opposite side of the lens from the source of illumination, and virtual focal points occur on the same side as the source of illumination.

Only one lens case is shown here, but the principles apply equally to all variations of lenses.

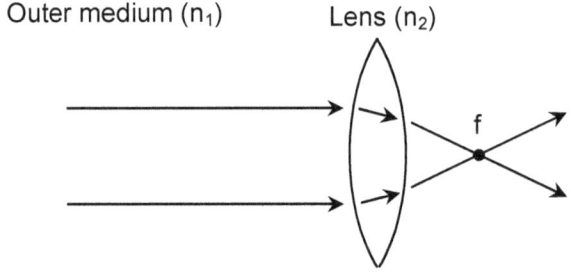

Care must be taken in properly identifying the normal and in applying Snell's law to both incidences of refraction for each ray.

Determining the point of image formation for mirrors and lenses is accomplished in a similar manner to that of the focal points. As with the focal points, image points may be either real or virtual, depending on the characteristics of the mirror or lens. To find the image point, two rays of differing angles must be traced as they interact with the lens or mirror. The initial directions of the rays can be chosen arbitrarily, but it is ideal to choose the directions such that the difficulty with determining the direction of the reflected or refracted ray is minimized. The intersection of these two rays is the image point. Only two examples are shown here, but the principles behind these examples may be applied to any variation of the situations, as well as to any combination of lenses and mirrors.

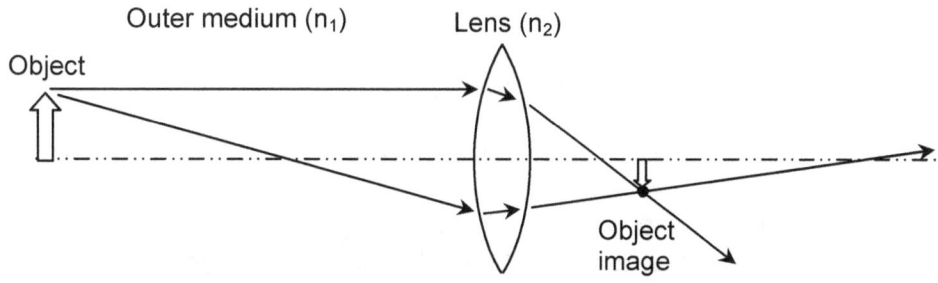

Skill 19.3 Applying the lens and mirror equations to analyze problems involving lenses and mirrors

Lenses
A thin lens in one in which focal length is much greater than lens thickness. For problems involving thin lenses, we can disregard any optical effects of the lens itself. Additionally, we can assume that the light that interacts with the lens makes a small angle with the optical axis of the system and so the sine and tangent values of the angle are approximately equal to the angle itself. This paraxial approximation, along with the thin lens assumptions, allows us to state:

$$\frac{1}{s} + \frac{1}{s'} = \frac{1}{f}$$

Where s=distance from the lens to the object (object location)
s'=distance from the lens to the image (image location)
f=focal length of the lens

Most lenses also cause some magnification of the object. Magnification is defined as:

$$m = \frac{y'}{y} = -\frac{s'}{s}$$

Where m=magnification
y'=image height
y=object height

Sign conventions will make it easier to understand thin lens problems:

Focal length: positive for a converging lens; negative for a diverging lens
Object location: positive when in front of the lens; negative when behind the lens
Image location: positive when behind the lens; negative when in front of the lens
Image height: positive when upright; negative when upside-down.
Magnification: positive for an erect, virtual image; negative for an inverted, real image

Problem:
A converging lens has a focal length of 10.00 cm and forms a 2.0 cm tall image of a 4.00 mm tall real object to the left of the lens. If the image is erect, is the image real or virtual? What are the locations of the object and the image?

Solution:
We begin by determining magnification:
$$m = \frac{y'}{y} = \frac{0.02m}{0.004m} = 5$$

Since the magnification is positive and the image is erect, we know the image must be virtual.

To find the locations of the object and image, we first relate them by using the magnification:
$$m = -\frac{s'}{s}$$
$$s' = -ms$$

Then we substitute into the thin lens equation, creating one variable in one unknown:
$$\frac{1}{s} + \frac{1}{s'} = \frac{1}{f}$$
$$\frac{1}{s} - \frac{1}{5s} = \frac{1}{10cm}$$
$$\frac{5-1}{5s} = \frac{1}{10cm}$$
$$s = \frac{40cm}{5} = 8cm \rightarrow\rightarrow s' = -5 \times 8cm = -40cm$$

Thus the object is located 8 cm to the left of the lens and the image is 40 cm to the left of the lens.

Mirrors
For image characteristics of curved mirrors see **Skill 19.1**. The relationship between the object distance from vertex s, the image distance from vertex s', and the focal length f is given by the equation

$$\frac{1}{s} + \frac{1}{s'} = \frac{1}{f}$$

TEACHER CERTIFICATION STUDY GUIDE

Magnification is defined as:

$$m = \frac{y'}{y} = -\frac{s'}{s}$$

where m=magnification, y'=image height, y=object height

Problem: A concave mirror collects light from a star. If the light rays converge at 50 cm, what is the radius of curvature of the mirror?

Solution: The point at which the rays converge is known as the focal point. The focal length, in this case, 50 cm, is the distance from the focal point to the mirror. The radius of curvature is the distance from the vertex to the center of curvature. The vertex is the point on the mirror where the principal axis meets the mirror. The center of curvature represents the point in the center of the sphere from which the mirror was sliced. Since the focal point is the midpoint of the line from the vertex to the center of curvature, or focal length, the focal length would be one-half the radius of curvature. Since the focal length in this case is 50 cm, the radius of curvature would be 100 cm.

Problem: An image of an object in a mirror is upright and reduced in size. In what type of mirror is this image being viewed, plane, concave, or convex?

Solution: The image in a plane mirror would be the same size as the object. The image in a concave mirror would be magnified if upright. Only a convex mirror would produce a reduced upright image of an object.

Skill 19.4   Applying Snell's law to analyze optical phenomena (e.g., total internal reflection, dispersion)

Snell's Law describes how light bends, or refracts, when traveling from one medium to the next. It is expressed as

$$n_1 \sin\theta_1 = n_2 \sin\theta_2$$

where n_i represents the index of refraction in medium i, and θ_i represents the angle the light makes with the normal in medium i.

Problem: The index of refraction for light traveling from air into an optical fiber is 1.44. (a) In which direction does the light bend? (b) What is the angle of refraction inside the fiber, if the angle of incidence on the end of the fiber is 22°?

PHYSICS

Solution: (a) The light will bend toward the normal since it is traveling from a rarer region (lower n) to a denser region (higher n).

(b) Let air be medium 1 and the optical fiber be medium 2:

$$n_1 \sin \theta_1 = n_2 \sin \theta_2$$
$$(1.00)\sin 22° = (1.44)\sin \theta_2$$
$$\sin \theta_2 = \frac{1.00}{1.44} \sin 22° = (.6944)(.3746) = 0.260$$
$$\theta_2 = \sin^{-1}(0.260) = 15°$$

The angle of refraction inside the fiber is $15°$.

Reflection may occur whenever a wave travels from a medium of a given refractive index to another medium with a different index. A certain fraction of the light is reflected from the interface and the remainder is refracted. However, when the wave is moving from a dense medium into one less dense, that is the refractive index of the first is greater than the second, a **critical angle** exists which will create a phenomenon known as **total internal reflection.** In this situation all of the wave is reflected. The critical angle of incidence θ_c is the one for which the angle of reflection is 90 degrees. Thus, according to Snell's law

$$n_1 \sin \theta_c = n_2 \sin 90° \Rightarrow \theta_c = \sin^{-1} \frac{n_2}{n_1}$$

Snell's law may also be used to understand the phenomenon of dispersion since it relates the angle of refraction at the boundary between two media to the relative refractive indices. Since different frequency components of visible light have different indices of refraction in any medium other than vacuum, each component of a beam of white light is refracted at a different angle when it crosses a surface resulting in a separation of the colors.

Skill 19.5 Analyzing given applications of lenses, mirrors, and prisms (e.g., telescopes, compound microscopes, eyeglasses)

Eye

The eye is a very complex sensory organ. Although there are many critical anatomical features of the eye, the lens and retina are the most important for focusing and sensing light. Light passes through the cornea and through the lens. The lens is attached to numerous muscles that contract and relax to move the lens in order to focus the light onto the retina. The pupil also contracts and relaxes to allow more or less light in the eye as required.

The retina contains rod cells which are responsible for vision in low light and cone cells which sense color and detail. Different types of cone cells are capable of sensing different wavelengths of light. A chemical called rhodopsin is present in the retina that converts light signals into electrical impulses that are sent to the brain to interpret as vision. The retina is lined with a black pigment called melanin that reduces reflection.

The eye relies on refraction to focus light onto the retina. Refraction occurs at four curved interfaces; between the air and the front of the cornea, the back of the cornea and the aqueous humor, the aqueous humor at the front of the lens, and the back of the lens and the vitreous humor. When each of these interfaces are working properly the light arrives at the retina in perfect focus for transmission to the brain as an image.

Eye Glasses
When all the parts of the eye are not working together correctly, corrective lenses or eyeglasses may be needed to assist the eye in focusing the light onto the retina. The surfaces of the lens or cornea may not be smooth causing the light to refract in the wrong direction. This is called astigmatism. Another common problem is that the lens is not able to change its curvature appropriately to match the image. The cornea can also be misshaped resulting in blurred vision. Corrective lenses consist of curves pieces of glass which bend the light in order to change the focal point of the light. A nearsighted eye forms images in front of the retina. To correct this, a minus lens consisting of two concave prisms is used to bend light out and move the image back to the retina. A farsighted eye creates images behind the retina. This is corrected using plus lenses that bend light in and bring the image forward onto the retina. The worse the vision, the farther out of focus the image is on the retina. Therefore the stronger the lens the further the focal point is moved to compensate.

Spectroscope
Spectrometers known as spectroscopes are used to identify materials. Spectroscopes are used often in astronomy and some branches of chemistry. Early spectroscopes were simply a prism with graduations marking wavelengths of light. Modern spectroscopes typically use a diffraction grating, a movable slit, and some kind of photo detector, all automated and controlled by a computer. When materials are heated they emit light that is characteristic of its atomic composition. The emission of certain frequencies of light produce a pattern of lines that are comparable to a fingerprint. The yellow light emission of heated sodium is a typical example.

A spectroscope is able to detect, measure and record the frequencies of the emitted light. This is done by passing the light though a slit to a collimating lens which transforms the light into parallel rays. The light is then passed through a prism that refracts the beam into a spectrum of different wavelengths. The image is then viewed alongside a scale to determine the characteristic wavelengths. Spectral analysis is an important tool for determining and analyzing the composition of unknown materials as well as for astronomical studies.

Camera

A camera is another device that utilizes the lens' ability to refract light to capture and process an image. As with the eye, light enters the lens of a camera and focuses the light on the other side. Instead of focusing on the retina, the image is focused on the film to create a film negative. This film negative is later processed with chemicals to create a photograph. A camera uses a converging or convex lens. This lens captures and directs light to a single point to create a real image on the surface of the film. To focus a camera on an image, the distance of the lens from the film is adjusted in order to ensure that the real image converges on the surface of the film and not in front of or behind it.

Different lenses are available which capture and bend the light to different degrees. A lens with more pronounced curvature will be able to bend the light more acutely causing the image to converge more closely to the lens. Conversely a flatter lens will have a longer focal distance. The further the lens is located from the film (flatter lens), the larger the image becomes. Thus zoom lenses on cameras are flat while wide angle lenses are more rounded. The focal length number on a certain lens conveys the magnification ability of the lens.

The film functions like the retina of the eye in that it is light sensitive and can capture light images when exposed. However, this exposure must be brief to capture the contrasting amounts of light and a clear image. The rest of the camera functions to precisely control how much light contacts the film. The aperture is the lens opening which can open and close to let in more or less light. The temporal length of light exposure is controlled by the shutter which can be set at different speeds depending on the amount of action and level of light available. The film speed refers to the size of the light sensitive grains on the surface of the film. The larger grains absorb more light photons than the smaller grains, so film speed should be selected according to lighting conditions.

Telescope

A telescope is a device that has the ability to make distant objects appear to be much closer. Most telescopes are one of two varieties, a refractor which uses lenses or a reflector which uses mirrors. Each accomplishes the same purpose but in totally different ways. The basic idea of a telescope is to collect as much light as possible, focus it, and then magnify it. The objective lens or primary mirror of a telescope brings the light from an object into focus. An eyepiece lens takes the focused light and "spreads it out" or magnifies it using the same principle as a magnifying glass using two curved surfaces to refract the light.

Microscope

Microscopes are used to view objects that are too small to be seen with the naked eye. A microscope usually has an objective lens that collects light from the sample and an eyepiece which brings the image into focus for the observer. It also has a light source to illuminate the sample. Typical optical microscopes achieve magnification of up to 1500 times.

SUBAREA V   **QUANTUM THEORY AND THE ATOM**

COMPETENCY 20.0   UNDERSTAND THE DUAL NATURE OF LIGHT AND MATTER

Skill 20.1   **Describing the quantization of energy in terms of Planck's theory**

Until Max Planck presented his theory of energy quantization late in the year 1900, energy was widely believed to consist of a continuous spectrum of values. The failure of this assumption to account for such phenomena as blackbody radiation led Planck to impose on his calculations the notion that energy could only take discrete values; that is, energy is quantized into "energy elements," or quanta.

Planck's theory indicated that the energy in a system is quantized and when the system changes energy states by emitting light, for example, the energy of the light must also have discrete energy values. The consequences of this principle are shown by its ability to account for the behavior of blackbody radiation but a more heuristically beneficial application is the relationship of the quantization of energy to the behavior of the atom.

An atom is composed of a positively charged, heavy nucleus surrounded by an electron "cloud." Since the electrons and the nucleus are oppositely charged, there is an attractive force that would, alone, appear to require that the electrons eventually collide with the nucleus, resulting in a collapse of the atom. From the study of electromagnetism, it is known that when a charged object is accelerated (as an electron would be in an orbit around the nucleus), it radiates energy in the form of electromagnetic waves. Nevertheless, if the possible energy levels of the atom are discrete instead of continuous, then it is feasible that the lowest level does not have sufficient energy to emit a quantum, thus preventing the electron from finally colliding with the nucleus. Therefore, the atom would be stable in spite of the presence of an electron that, since it is still "orbiting," would otherwise be expected to radiate and continually lose energy.

Quantization of energy thus supports atomic theory whereas a purely classical approach involving continuous energy levels would result in a catastrophic failure of the theory. The energy emitted from an atom in the form of light is calculated as the product of the frequency and Planck's constant h. Since h is approximately 6.626×10^{-34} Joule · second, a virtually infinitesimal number, the energies involved at the atomic level are so tiny that they have no apparent impact on the macroscopic world. As a result, quantization effects are not noticeable for macroscopic objects but only become apparent for objects like atoms and subatomic particles.

Planck's theory also leads to the notion that quanta of light, rather than being purely continuous waves as they may appear to be in everyday life, are instead more like "packets" of energy. The concept of these energy packets, or photons, was essential to a more empirically useful understanding of the photoelectric effect, as presented in 1905 by Albert Einstein. Light was then understood as having the characteristics of particles in addition to the characteristics of waves. This concept was also crucial to the development of quantum mechanics in subsequent decades.

Skill 20.2 Applying the laws of photoelectric emission to explain photoelectric phenomena

Einstein used the photon hypothesis to explain the photoelectric effect which is the **emission of electrons from a metal surface when light is incident on it**. When this metal surface is a cathode with the anode held at a higher potential V, the emitted electrons create a current flow in the external circuit. It is observed that current flows only for light of higher frequencies, i.e. electrons are released from the metal only by high frequency light. Also there is a threshold negative potential, the **stopping potential** V_0 below which no current will flow in the circuit.

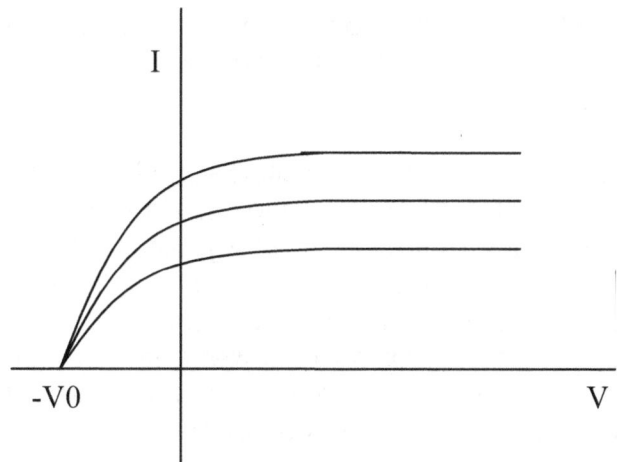

The figure displayed above shows current flow vs. potential for three different intensities of light. It shows that the maximum current flow increases with increasing light intensity but the stopping potential remains the same.

All these observations are counter-intuitive if one considers light to be a wave but may be understood in terms of light particles or photons. According to this interpretation, each photon transfers its energy to a single electron in the metal. Since the energy of a photon depends on its frequency, only a photon of higher frequency can transfer enough energy to an electron to enable it to pass the stopping potential threshold.

When V is negative, only electrons with a kinetic energy greater than $|eV|$ can reach the anode. The maximum kinetic energy of the emitted electrons is given by eV_0. This is expressed by Einstein's photoelectric equation as

$$(\tfrac{1}{2}mv^2)_{max} = eV_0 = hf - \varphi$$

where the **work function** φ is the energy needed to release an electron from the metal and is characteristic of the metal.

Problem: The work function for potassium is 2.20eV. What is the stopping potential for light of wavelength 400nm?

Solution:
$eV_0 = hf - \varphi = hc/\lambda - \varphi = 4.136 \times 10^{-15} \times 3 \times 10^8 / (400 \times 10^{-9}) - 2.20 = 3.10 - 2.20 = 0.90\text{eV}$
Thus stopping potential $V_0 = 0.90\text{V}$

Skill 20.3 Applying the principles of stimulated emission of radiation to lasers and masers

The names "laser" and "maser" are acronyms for "light amplification by stimulated emission of radiation" and "microwave amplification by stimulated emission of radiation." Thus, as is evident by their names, these two devices are based on the principle of stimulated emission.

According to quantum theory, an atom in an excited state has a certain probability in a given time frame for relaxing to a lower state through, for example, the emission of a photon of the same energy as the energy difference between the two states. Stimulated emission, however, may take place when the excited atom is perturbed by a passing photon of the same energy as the excitation. In such a case, the atom relaxes to a lower energy by emitting a new photon, resulting in a total of two photons of equal frequency (and, thus, energy) and equal phase. That is to say, the photons are coherent.

In order to produce a significant level of stimulated emission for application in typical lasers and masers, population inversion is required. Population inversion is a situation in which there are more atoms in a particular excited state than in a particular lower-energy state. In order to achieve this condition, the material must be "pumped," which can be performed using electromagnetic fields (light). Pumping to produce population inversion often requires three or more atomic energy levels.

Higher energy level (E_3)

The above energy level diagram is for a so-called three-level laser (or maser). Other lasers may use more levels, depending on the material that is being pumped and the desired frequency output. The transition between the energy levels E_3 and E_2 is noted as being fast and radiationless; that is, almost as soon as the electron(s) of an atom are pumped to E_3, they decay to E_2 without radiating light. The energy difference can be emitted in the form of a phonon (vibration mode in the material, or heat).

The abundance of excited atoms due to population inversion allow for a "chain reaction" to form when photons of the proper frequency (proportional by Planck's constant to the energy difference between the excited state and lower-energy state) are incident. This results in a cascading increase in photon intensity through stimulated emission, thus producing the coherent, high-power light that is used by the laser or maser. This process has its limits, of course, as determined by how quickly the atoms can be pumped in relation to the relaxation processes. A saturation level exists beyond which the rate of stimulated emission cannot be increased. Spontaneously emitted photons can start the cascading process of stimulated emission.

The main components of a laser are a gain medium, an energy source (pump) for the gain medium and a resonant optical cavity with a partial transparency in one mirror. The resonant cavity is tuned to a particular frequency such that the photons of the desired laser frequency are coherent and, largely, isolated within the cavity. A partially transparent mirror on one end of the cavity allows some of the light to exit, thus producing the laser beam.

Masers, which operate at lower frequencies than lasers, generally rely on the same principles as lasers, although the types of gain media and resonant cavities may differ.

Skill 20.4 Analyzing evidence supporting the dual nature of light and matter

Scientists have argued for years whether light is a wave or a stream of particles. Actually, light exhibits the behaviors of both waves and particles.

Wave characteristics
Light undergoes **reflection**, **refraction**, and **diffraction** just as any wave would. The image you see in a mirrored surface is the result of the reflection of the light waves off the surface. Light waves follow the "law of reflection," i.e. the angle at which the light wave approaches a flat reflecting surface is equal to the angle at which it leaves the surface. When light crosses the boundary between two different media, its path is bent, or refracted. Diffraction occurs when light encounters an obstacle in its path or passes through an opening. Light diffracts around the sides of an object causing the shadow of the object to appear fuzzy.

Another phenomenon unique to waves is **wave interference**. This characteristic describes what happens when two waves meet while traveling along the same medium. If light **constructively interferes** (trough meets trough or crest meets crest), the two light waves reinforce one another to produce a stronger light wave. However, if light **destructively interferes** (crest meets trough), the two light waves destroy each other and no light wave is produced.

Polarization changes unpolarized light into polarized light. This process can only occur with a transverse wave. An everyday example of polarization is found in polarized sunglasses which reduce glare.

Particle characteristics
Einstein came up with the quantum theory of light that states that light is made up of **photons** or **quanta**, discrete particles of electromagnetic radiation. The photons, or individual particles of light, have been shown to have isolated arrival times. A movie was taken of the comet Hyakutake that showed a breakdown of the photons traveling with the comet and scattered throughout the region. Some phenomena such as **blackbody radiation** and the **photoelectric effect** can only be explained using the particle nature of light.

Presently, a combination of the two theories, or **wave - particle duality,** is accepted.

Skill 20.5 Solving problems using de Broglie's equation

Following the concept of wave-particle duality for light (photons) as introduced by Max Planck through his theory of quantization of energy and by Albert Einstein through his explanation of the photoelectric effect, Louis de Broglie hypothesized that matter also has the characteristics of both particles and waves. The de Broglie wavelength λ can be related to the momentum p and Planck's constant h by way of the following equation.

$$\lambda = \frac{h}{p}$$

The momentum of the particle, p, is defined as p = mv, where m is the relativistic mass (especially if the particle has a velocity near the speed of light). Also, based on the Planck relationship of energy and wavelength for a quantum (E = hv), the de Broglie equation for the frequency v of a particle may be derived. Here, E is the kinetic energy of the particle.

$$v = \frac{E}{h}$$

Given that Planck's constant is extremely small (6.626×10^{-34} Joule·second), the de Broglie wavelength is also vanishingly small for macroscopic objects. On the microscopic scale, however, subatomic particles have sufficiently small momenta to yield significant de Broglie wavelengths; thus, while this equation has little apparent relevance to macroscopic objects, it is useful for dealing with quantum mechanical problems on a microscopic scale.

Simple problems typically involve finding the de Broglie wavelength of subatomic particles such as electrons. An example problem is given below.

Problem: Find the wavelength of an electron traveling at 6.7×10^6 meters per second.

Solution: Use de Broglie's equation, incorporating the mass of an electron (9.11×10^{-31} kilograms). The speed of the particle is sufficiently low to justify neglecting relativistic effects.

$$\lambda = \frac{6.626 \times 10^{-34} \, J \cdot s}{9.11 \times 10^{-31} \, kg \cdot 6.7 \times 10^6 \, m/s} = 0.109 \text{ nm}$$

The de Broglie wavelength can have relevance to empirical situations and calculations, as demonstrated by the Davisson-Germer experiment. In this experiment, a beam of electrons with known momenta is incident on a nickel target. The scattering of the electron beam shows a diffraction pattern consistent with a wave of the corresponding de Broglie wavelength. This result indicates the validity of de Broglie's equation and shows it to have apparent relevance to actual particles.

Furthermore, de Broglie's hypothesis, which suggests that particles also have wave-like behavior, helped to solidify the foundation of quantum theory which is based on Schrödinger's wave equation as a mathematical expression of the behavior of particles.

Skill 20.6 Applying the uncertainty principle in various situations

Heisenberg's uncertainty principle places a limit on the accuracy with which one can measure the properties of a physical system. This limit is not due to the imperfections of measuring instruments or experimental methods but arises from the fundamental wave-particle duality inherent in quantum systems.

One statement of the uncertainty principle is made in terms of the position and momentum of a particle. If Δx is the uncertainty in the position of a particle in one dimension and Δp the uncertainty in its momentum in that dimension, then according to the uncertainty principle
$$\Delta x \Delta p \geq \hbar/2$$
where the reduced Planck's constant $\hbar = h/2\pi = 1.05457168 \times 10^{-34} J.s$

Thus if we measure the position of a particle with greater and greater accuracy, at some point the accuracy in the measurement of its momentum will begin to fall. A simple way to understand this is by considering the wave nature of a subatomic particle. If the wave has a single wavelength, then the momentum of the particle is also exactly known using the DeBroglie momentum-wavelength relationship. The position of the wave, however, extends through all space. If waves of several different wavelengths are superposed, the position of the wave becomes increasingly localized as more wavelengths are added. The increased spread in wavelength, however, then results in an increased momentum spread.

An alternate statement of the uncertainty principle may be made in terms of energy and time.
$$\Delta E \Delta t \geq \hbar/2$$

Thus, for a particle that has a very short lifetime, the uncertainty in the determination of its energy will be large.

Problem: If a proton is confined to a nucleus that is approximately $10^{-15} m$ in diameter, estimate the minimum uncertainty in its momentum in any one dimension.

Solution: The uncertainty of the position of the proton in any dimension cannot be greater than $10^{-15} m$. Using the uncertainty principle we find that the approximate uncertainty in its momentum in any one dimension must be greater than $\Delta p = \hbar / (2\Delta x) \approx 10^{-19} Kg.m/s$

TEACHER CERTIFICATION STUDY GUIDE

COMPETENCY 21.0   UNDERSTAND PHYSICAL MODELS OF ATOMIC STRUCTURE AND THE NATURE OF ELEMENTARY PARTICLES

Skill 21.1   Analyzing historic and contemporary models of atomic structure (e.g., Rutherford, Bohr, Schrödinger, Dirac, Heisenberg, Pauli)

In the West, the Greek philosophers Democritus and Leucippus first suggested the concept of the atom. They believed that all atoms were made of the same material but that varied sizes and shapes of atoms resulted in the varied properties of different materials. By the 19th century, John Dalton had advanced a theory stating that each element possesses atoms of a unique type. These atoms were also thought to be the smallest pieces of matter which could not be split or destroyed.

Atomic structure began to be better understood when, in 1897, JJ Thompson discovered the electron while working with cathode ray tubes. Thompson realized the negatively charged electrons were subatomic particles and formulated the "plum pudding model" of the atom to explain how the atom could still have a neutral charge overall. In this model, the negatively charged electrons were randomly present and free to move within a soup or cloud of positive charge. Thompson likened this to the dried fruit that is distributed within the English dessert plum pudding though the electrons were free to move in his model. Ernest Rutherford disproved this model with the discovery of the nucleus in 1909. Rutherford proposed a new "planetary" model of the atom in which electrons orbited around a positively charged nucleus like planets around the sun. Over the next 20 years, protons and neutrons (subnuclear particles) were discovered while additional experiments showed the inadequacy of the planetary model.

As quantum theory was developed and popularized (primarily by Max Planck and Albert Einstein), chemists and physicists began to consider how it might apply to atomic structure. Niels Bohr put forward a model of the atom in which electrons could only orbit the nucleus in circular orbitals with specific distances from the nucleus, energy levels, and angular momentums. In this model, electrons could only make instantaneous "quantum leaps" between the fixed energy levels of the various orbitals. The Bohr model of the atom was altered slightly by Arnold Sommerfeld in 1916 to reflect the fact that the orbitals were elliptical instead of round.

Though the Bohr model is still thought to be largely correct, it was discovered that electrons do not truly occupy neat, cleanly defined orbitals. Rather, they exist as more of an "electron cloud." The work of Louis de Broglie, Erwin Schrödinger, and Werner Heisenberg showed that an electron can actually be located at any distance from the nucleus. However, we can find the *probability* that the electrons exists at given energy levels (i.e., in particular orbitals) and those probabilities will show that the electrons are most frequently organized within the orbitals originally described in the Bohr model.

Skill 21.2   Interpreting notation used to represent elements, molecules, ions, and isotopes

Every element in the periodic table is represented by its own symbol consisting of one or more letters. (E.g. H for hydrogen, He for Helium, O for oxygen). In general, a molecule of an element or a compound is represented by a formula that shows what atoms and how many of each constitute the molecule. For instance, water consists of two hydrogen and one oxygen atom. Thus a molecule of water is represented as H_2O. This is the kind of notation used in chemical equations.

Atoms and ions of a given element that differ in number of neutrons have a different mass and are called **isotopes**. In writing nuclear equations, where isotopes of the same element must be distinguished, it is helpful to use a notation where the number of nucleons and protons/electrons is included along with the element symbol. The identity of an **element** depends on the **number of protons** in the nucleus of the atom. This value is called the **atomic number** and it is sometimes written as a subscript before the symbol for the corresponding element. A nucleus with a specified number of protons and neutrons is called a **nuclide**, and a nuclear particle, either a proton or neutron, may be called a **nucleon**. The total number of nucleons is called the **mass number** and may be written as a superscript before the atomic symbol.

$^{14}_{6}C$ represents an atom of carbon with 6 protons and 8 neutrons.

The **number of neutrons** may be found by **subtracting the atomic number from the mass number**. For example, uranium- 235 has 235−92=143 neutrons because it has 235 nucleons and 92 protons.

An ion is an atom or molecule with a net positive or negative charge (due to an unequal number of protons and electrons) and is represented with a plus or minus sign and the number of excess electrons or protons placed at the top right-hand corner. For example, a positively charged sodium ion with one extra proton may be represented as Na^{+1}.

Skill 21.2   Analyzing bright-line spectra in terms of electron transitions

Quantum #	Radius
$n \to \infty$	$r_\infty \to \infty$
⋮	⋮
$n = 5$	$r_5 = 25a_0$
$n = 4$	$r_4 = 16a_0$
$n = 3$	$r_3 = 9a_0$
$n = 2$	$r_2 = 4a_0$
$n = 1$ ⊕ (H nucleus)	$r_1 = a_0$

An electron may exist at distinct radial distances (r_n) from the nucleus. These distances are proportional to the square of the **principal quantum number**, n. For a hydrogen atom (shown at left), the proportionality constant is called the **Bohr radius** ($a_0 = 5.29 \times 10^{-11}$ m). This value is the mean distance of an electron from the nucleus at the ground state of $n = 1$. The distances of other electron shells are found by the formula:

$$r_n = a_0 n^2.$$

As $n \to \infty$, the electron is no longer part of the hydrogen atom. Ionization occurs and the atom become an H^+ ion.

A quantum of energy (ΔE) emitted from or absorbed by an electron transition is directly proportional to the frequency of radiation. The proportionality constant between them is **Planck's constant** ($h = 6.63 \times 10^{-34}$ J·s):

$$\Delta E = h\nu \quad \text{and} \quad \Delta E = \frac{hc}{\lambda}.$$

The energy of an electron (E_n) is inversely proportional to its radius from the nucleus. For a hydrogen atom, only the principle quantum number determines the energy of an electron by the **Rydberg constant** ($R_H = 2.18 \times 10^{-18}$ J):

$$E_n = -\frac{R_H}{n^2}.$$

The Rydberg constant is used to determine the energy of a photon emitted or absorbed by an electron transition from one shell to another in the H atom:

$$\Delta E = R_H \left(\frac{1}{n_{initial}^2} - \frac{1}{n_{final}^2} \right).$$

PHYSICS

When a photon is absorbed, n_{final} is greater than $n_{initial}$, resulting in positive values corresponding to an endothermic process. Ionization occurs when sufficient energy is added for the atom to lose its electron from the ground state. This corresponds to an electron transition from $n_{initial} = 1$ to $n_{final} \to \infty$. The Rydberg constant is the energy required to ionize one atom of hydrogen. Photon emission causes negative values corresponding to an exothermic process because $n_{initial}$ is greater than n_{final}.

Planck's constant and the speed of light are often used to express the Rydberg constant in units of s^{-1} or length. The formulas below determine the photon frequency or wavelength corresponding to a given electron transition:

$$v_{photon} = \left(\frac{R_H}{h}\right)\left|\frac{1}{n_{initial}^2} - \frac{1}{n_{final}^2}\right| \quad \text{and} \quad \lambda_{photon} = \frac{1}{\left(\frac{R_H}{hc}\right)\left|\frac{1}{n_{initial}^2} - \frac{1}{n_{final}^2}\right|}.$$

These formulas **relate observed lines in the hydrogen spectrum to individual transitions** from one quantum state to another.

A simple optical spectroscope separates visible light into distinct wavelengths by passing the light through a prism or diffraction grating. When electrons in hydrogen gas are excited inside a discharge tube, the emission spectroscope shown below detects photons at four visible wavelengths.

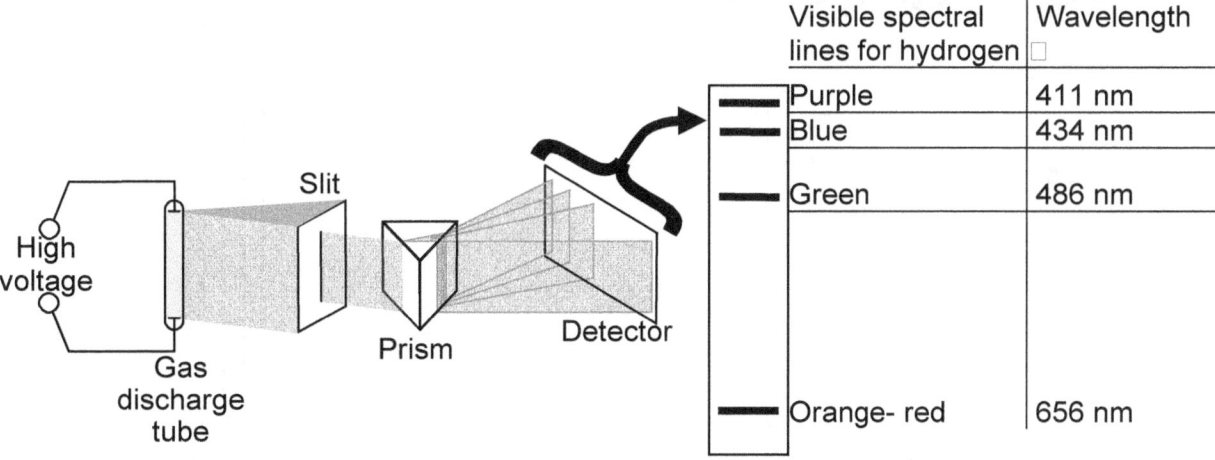

Visible spectral lines for hydrogen	Wavelength
Purple	411 nm
Blue	434 nm
Green	486 nm
Orange-red	656 nm

Every line in the hydrogen spectrum corresponds to a transition between electron energy levels. The four spectral lines from hydrogen emission spectroscopy in the visible range correspond to electron transitions from n = 3, 4, 5, and 6 to n =2 as shown in the table below.

Radiation type	Wavelength ×(nm)	Frequency ×(s^{-1})	Energy change ×E (J)	Electron transition $n_{initial} \rightarrow n_{final}$
Ultraviolet	≤397	≥7.55×10^{14}	≤−5.00×10^{-19}	∞→1, … 2→1 ∞→2, … 7→2
Purple	411	7.31×10^{14}	−4.84×10^{-19}	6→2
Blue	434	6.90×10^{14}	−4.58×10^{-19}	5→2
Green	486	6.17×10^{14}	−4.09×10^{-19}	4→2
Orange-red	656	4.57×10^{14}	−3.03×10^{-19}	3→2
Infrared and beyond	≥821	≤3.65×10^{14}	≥−2.42×10^{-19}	∞→3, … 4→3 ∞→4, … 5→4 ⋮

Most lines in the hydrogen spectrum are not at visible wavelengths. Larger energy transitions produce ultraviolet radiation and smaller energy transitions produce infrared or longer wavelengths of radiation. Transitions between the first three and the first six energy levels of the hydrogen atom are shown in the diagram to the right. The energy transitions producing the four visible spectral lines are colored grey.

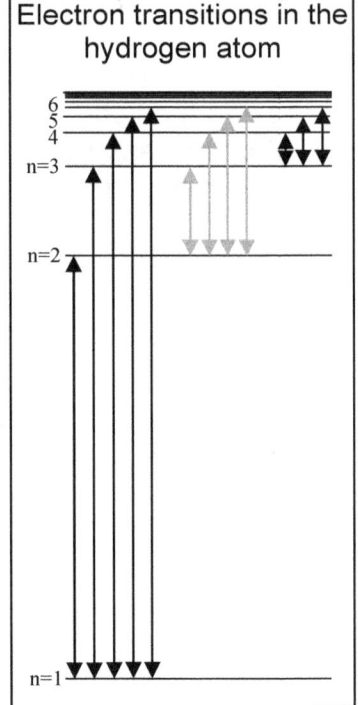

Electron transitions in the hydrogen atom

Skill 21.3 Recognizing the relationship between the design of particle accelerators and elementary particle characteristics

Elementary (fundamental) particles are, in many ways, theoretical. These particles are not seen with the unaided (or aided) human eye, and therefore must be understood based on other phenomena that are, presumably, associated with or caused by their presence or action. As such, experimentally investigating fundamental particles requires a theoretical framework for design of the experimental apparatus and for interpretation of the results. With the recent advent of inexpensive computing power, complex and repetitive calculations can be conducted quickly and relatively inexpensively, but the numbers that are output by the computer are only as good as the data that is input and the interpretation that is later applied to the results.

First, it is critical to note that **the "resolution" of an experiment involving fundamental particles corresponds very closely to the kinetic energy of the particles involved**. As with light (photons), the wavelength of the illuminating particles must be smaller than the smallest desired dimension of resolution in order to "see" the object of interest. As such, a particle accelerator must be designed with the dimensions of the proposed investigation clearly in mind. That is to say, the energy capabilities of the accelerator must meet or exceed the requirements for a given resolution (such as, for example, investigation of quarks).

The method for accelerating the particles is key; fundamental particles, of course, cannot be accelerated reliably by mechanical or conventional means. The use of electric and magnetic fields and waves allows for the acceleration of particles, but only charged particles. Thus, if neutral particles are to be investigated in this manner, their production must be possible using charged particles (e.g., by way of charged particle collisions). The characteristics of the particles to be investigated using the accelerator must then be considered carefully and integrated into the design of the equipment.

Some form of target must be used to facilitate the production of collision events and to allow study of the particles. Using the analogy of light shining on an object, fundamental particles are used to "illuminate" an object (perhaps another fundamental particle) to examine its behavior and structure. The target may simply consist of another beam of particles traveling in the opposite direction, forcing the particles to collide. Alternatively, some fixed target may be used and may contain any number of materials, depending on the experiments of interest.

Last, **detection and transduction of data must be achieved through appropriate design of detectors and other equipment**. This equipment is responsible for converting an event associated with the detection of a particle into a signal recognizable by a computer and, ultimately, analyzable by the experimenters. Of interest in the design process is, for example, the charge of the particles that originate from a collision event. These particles can be deflected by electric or magnetic fields, and the associated detection equipment can take advantage of these properties. Photons, an example of an uncharged particle, can be detected with photomultiplier tubes. Thus, the types of detection equipment necessary in the design of an accelerator depend on the characteristics of the particles of interest.

The four main points in the design of an accelerator include the particle kinetic energy (and, concomitantly, the resolution) necessary for the proposed experiments, the form of particle accelerator appropriate to the particular particles, the target to be used for producing collision events and the detection equipment necessary for the specific particles being investigated. Of course, proper computer handling of the data is likewise crucial. Since computers simply output numbers, a proper theory for interpretation is necessary.

COMPETENCY 22.0 UNDERSTAND THE STANDARD MODEL OF PARTICLE PHYSICS

Skill 22.1 Recognizing key contributions by individuals (e.g., Gell-Mann, Feynman) in the development of the standard model

The Standard Model of particle physics is a quantum field theory that combines quantum mechanics and special relativity in an attempt to explain the behavior of fundamental particles. Like most complex theories, the Standard Model originated from the contributions of numerous individuals over decades, and several of these individuals are noteworthy for their key work. A full history of the development of quantum mechanics and special relativity would be necessary to obtain a complete picture of the origins of the Standard Model, but it is helpful to review a few of the main efforts specifically relevant to this particular physical theory.

Murray Gell-Mann and George Zweig proposed the existence of quarks in 1964. This development eventually led to the quark model used in the Standard Model, which includes up, down, charm, strange, top and bottom quarks, each of which is defined by its associated quantum numbers. Along with electrons, muons and tau particles, as well as the associated three types of neutrinos, quarks compose the group of particles called fermions. Although the theory of Gell-Mann and Zweig was incomplete at the time, it formed a key basis for further development the Standard Model. **Gell-Mann, along with Harold Fritzsch, also introduced the concept of "color" charge**, which constituted the foundation for an understanding of the strong nuclear force through **quantum** chromodynamics (QCD).

In 1967, Steven Weinberg and Sheldon Glashow developed the electroweak theory, which presented a unified account of the electromagnetic and weak nuclear forces. This effort was paralleled by **Abdus Salam**, who is also credited with having independently developed electroweak theory simultaneously. The work of these individuals led to a Nobel Prize in physics in 1979.

At the end of the 1960s, **Jerome Friedman, Henry Kendall and Richard Taylor**, who worked at the Stanford Linear Accelerator Center (SLAC), experimentally **uncovered the first evidence for the actual existence of quarks**. The three scientists were jointly awarded the Nobel Prize in physics in 1990 for their work in the area of electron scattering, which led to a greater experimental understanding of quark theory.

Richard Feynman, along with Julian Schwinger and Shin-Ichiro Tomonaga, developed quantum electrodynamics (QED), leading to a Nobel Prize in physics in 1965. QED provides a fundamental theory describing the interaction of protons, neutrons and electrons, and was essential for a more complete understanding of modern particle physics. Feynman is also known for creating the so-called Feynman diagrams that depict the interactions of fundamental particles.

The Standard Model was presented in summary fashion by John Iliopoulos in a conference report. This report combined the details of the preceding work by other scientists in developing the Standard Model and was the first cohesive presentation of the Model.

Skill 22.2 Describing historical developments leading to the standard model

A number of critical historical events in the world of physics, especially in the early half of the 20th century, led to the development of the Standard Model. **Of fundamental importance were the development of special relativity and the development of quantum mechanics**; these concepts were fused to build the understanding of particle physics that underlies the Standard Model.

The theory of special relativity was proposed in 1905 by Albert Einstein in a work entitled "On the Electrodynamics of Moving Bodies." This theory serves a vital role in high-energy physics, especially with regard to the design and understanding of particle accelerators and other experimental equipment that have been used to investigate quarks and other aspects of the Standard Model.

Quantum mechanics arose out of the combined efforts of a number of individuals. One of the motivations for the move from a more classical approach to the quantum mechanical approach was the failure of classical mechanics to explain various phenomena including blackbody radiation and the spectrum of the hydrogen atom. In the same year as his composition on special relativity, Einstein presented an explanation of the photoelectric effect that envisioned light as having the characteristics of particles as well as waves. This concept was extended by Louis de Broglie in the mid 1920s to include the wave-particle duality of matter, in addition to that of light.

In the mid 1910s, Niels Bohr refined the Rutherford model of the atom to include discrete electron orbits, rather than a continuous spectrum, thus accounting for the discrete "lines" in the radiation spectrum of hydrogen. This model included the notion of radiation in terms of quantized packets of energy, or "quanta," rather than a continuous range of possible energies. This idea tied into the wave-particle duality of light suggested by Einstein in his investigation of the photoelectric effect.

Arguably the most definitive development in the history of quantum mechanics came with Erwin Schrödinger's wave equation which presented a unified mathematical conception of the behavior of particles. A formulation of quantum mechanics was then made possible and its development realized in the mid 1920s. Combined with the uncertainty principle of Werner Heisenberg, the basic theoretical machinery was in place for the development of the Standard Model of physics.

Theoretical breakthroughs more directly relevant to the Standard Model included the **development of quantum electrodynamics (QED) by Richard Feynman, Julian Schwinger and Shin-Ichiro Tomonaga**, as well as the **development of quantum chromodynamics (QCD) by Murray Gell-Mann and Harold Fritzsch.** QED presented a more fundamental understanding of the interaction of electrically charged particles than did classical electrodynamics, and QCD provided an account of the strong nuclear force by way of "color" charge.

The work of Gell-Mann and George Zweig in developing the theory of quarks was given experimental backing in the late 1960s by Jerome Friedman, Henry Kendall and Richard Taylor. The work of these latter three at the Stanford Linear Accelerator Center (SLAC) yielded the first experimental evidence of the existence of quarks. Quarks, along with electrons, muons, tau particles and the three types of neutrinos associated with these leptons, became the basis for the set of fundamental particles that compose matter (fermions) according to the Standard Model.

The neutrino was postulated in 1930 by Wolfgang Pauli, but experimental evidence indicating the discovery of a neutrino was not reported until 1956, when Frederick Reines (who won the Nobel Prize in 1995 for this work) and others published claims of having actually detected this electrically neutral particle.

Over time, since the first full presentation of the Standard Model by John Iliopoulos in 1974, experimental evidence has been reported for detection of various other particles, such as the various quarks and bosons, described in the Model. The Standard Model, however, has mostly been the outworking of a synthesis of quantum mechanics and special relativity, as applied to the search for the basic constituents of matter.

Skill 22.3 Describing experimental techniques and methods used to investigate elementary particles

The investigation of elementary (or fundamental) particles is based largely upon the outworking of quantum mechanics and special relativity. Some subsidiary theories along these lines include quantum electrodynamics (QED), which describes the behavior of electrically charged particles, and quantum chromodynamics (QCD), which describes the strong nuclear force through the "color" charge. Given that the fundamental particles being sought are not discernable by the unaided (or aided) human eye, the experimental techniques and methods used to investigate these particles are largely based on the theory that describes them.

Since individual fundamental particles are not apparent in everyday life (or even in many more complicated situations), special equipment for the investigation of these particles must be used. **Particle accelerators of various types are the most useful, and recognizable, apparatus for high-energy (particle) physics**. Particle physics often requires high-energy accelerators, since, in order to decrease the wavelength of the particle (and thereby increase the "resolution" of the experiment), the momentum (and, consequently, the kinetic energy) of the particle must be increased. According to the theory, the use of high-energy particles allows for a more clear understanding of the behavior and structure of the particles. This situation is analogous to the inability of light to resolve objects smaller than the wavelength of the light; thus, similarly, better resolution of the structure and behavior of particles can only be achieved through the use of increased kinetic energy.

Particle accelerators seek to increase the resolution of the experiment by increasing the speed of the particles, thereby decreasing their associated wavelength. Collision events can be produced by using a particle beam and a fixed target or two particle beams that are coincidental, allowing particles in each beam to collide (thus yielding higher relative particle energies than fixed target experiments). Particles can be accelerated by way of electric and magnetic fields that keep the beam confined and increase the kinetic energy. Collisions between particles or between particles and some target (i.e., "events") yield a "debris" of scattered or newly formed particles, each with a charge, momentum and direction of propagation that can provide insight into the behavior and, in some cases, internal structure of the original particles.

Surrounding the fixed target or specific beam location are detectors that are designed specifically for various types of particles. Detection of charged particles can take advantage of the positive or negative charge while detectors for various neutral particles must take different approaches.

In addition to the highly expensive and complex particle accelerators, simpler experiments can also be conducted, as took place prior to the advent of the accelerator. Electron beams can be formed, for example, in the ubiquitous cathode ray tube. Detection of charged particles can be achieved through the use of the bubble chamber which involves a transparent liquid and a magnetic field. Charged particles leave trails through the liquid and the magnetic field causes deflection as the particles travel through it. The bubble chamber, although having been largely replaced by the more complicated particle accelerators and their associated detection equipment, also provided a method for studying particle physics. The cloud chamber was similar in nature, using a gas instead of a liquid. (Also see **Skill 21.3**).

Skill 22.4 Describing the families of subnuclear particles (e.g., fermions, bosons, baryons, hadrons) and their properties

There are two types of elementary particles: **fermions** and **bosons**. Bosons have integer spin, while fermions have half integer spin.

While there are many subatomic bosons, the most familiar one is the **photon**. A photon has zero mass and charge; in a vacuum, a photon travels at the speed of light. Photons do not spontaneously decay, but can be emitted or absorbed by atoms via a number of natural processes. In fact, photons compose all forms of light and mediate electromagnetic interactions.

The two basic types of fermions are quarks and leptons.

Quarks: Quarks are found nearly exclusively as components of neutrons and protons. They come in three types of arbitrarily named flavors: up, charm, and top (which have a charge of +2/3) and down, strange, and bottom (which have a charge of –1/3. The flavors have varying mass which must be found via indirect methods.

Leptons: Unlike quarks, leptons do not experience strong nuclear force. There are three flavors of leptons; the muon, the tau, and the electron. Each type of lepton consists of a massive charged particle with the same name as the flavor and a smaller neutral particle. This nearly massless neutral particle is called a **neutrino**. Each lepton has a charge of +1 or –1.

For hadrons and baryons see next section.

TEACHER CERTIFICATION STUDY GUIDE

Skill 22.5 Explaining the properties of quarks and how they interact to form protons, neutrons, and other particles

Quarks and leptons were introduced in the previous section. Here we analyze how they combine to form the subatomic particles listed below. In each case, we can compare charge, make-up, mass, location, and mobility of the particle.

Electrons: Electrons are a specific type of lepton that exist outside of the nucleus in positions that are functions of their energy levels. Their arrangement and interactions are key to all chemical bonding and reactions. During such chemical processes, electrons can easily be transferred from one atom to another. The charge of an electron is defined as −1 in atomic units (actual charge is -1.6022×10^{-19} coulomb) and the mass of an electron is $1/1836$ of that of a proton.

Protons: Protons are confined inside the atomic nucleus and have a defined charge of +1 atomic units (1.602×10^{-19} coulomb). The mass of a proton is 1 Dalton (or 1 atomic mass unit). Protons may be lost from the nucleus in certain types of radioactive decay. A proton is composed of 2 up quarks and one down quark.

Neutrons: Like protons, neutrons are confined to the nucleus and only lost during certain types of radioactive decay. Neutrons are uncharged particles with assigned mass of 1 Da (in reality neutrons are slightly heavier than protons). A neutron is composed of 2 down quarks and one up quark.

Heavy particles, such as protons and neutrons, that interact by strong interactions are called **hadrons**. **Baryons** are hadrons that are made up of three quarks and are fermions (due to odd number of quarks). Protons and neutrons as well as lambda, sigma, xi and omega particles are baryons. **Mesons** are hadrons that are made up of two quarks and are bosons (due to even number of quarks).

Skill 22.6 Describing relationships between particles and antiparticles

According to quantum mechanics, a particle and its corresponding antiparticle both have the same mass and the same quantum spin state, but opposite charge. Thus, an electron and its antimatter counterpart, a positron, are identical except that the positron has an opposite charge of equal magnitude to that of the electron. The existence of antiparticles was first theorized by Paul Dirac in accord with the implications of the so-called Dirac equation.

Although the difference in charge is the only apparent difference between particles and antiparticles, neutral particles are not necessarily the same as their corresponding antiparticles.

The neutron, for example, is, as an aggregate particle, neutral. Nevertheless, its quark substructure, which includes one up quark and two down quarks, differs between the neutron and the antineutron. In the case of the antineutron, an anti-up and two antidown quarks compose the substructure, each of these having an opposite charge to that of the respective up and down quarks. Some neutral bosons, such as photons, are indistinguishable from their respective antiparticles, however.

A property of particle-antiparticle pairs is that they can be annihilated to form other particles. One of the most commonly cited examples of this phenomenon is the annihilation of an electron and positron to form photons with equivalent energy. Likewise, in a reverse manner, photons of sufficient energy can create an electron-positron pair. The energy contained in the photons must, for this to take place, be at least equal to the energy contained in the rest mass of the electron and positron. Energy above this level would be converted into kinetic energy. Momentum and energy are conserved in these reactions. In accordance with special relativity, an integral part of the Standard Model of particle physics, the relativistic mass of the particles must be considered when performing analyses of particle interactions.

The close relationship between particles and antiparticles has led to an unsolved conundrum regarding the material composition of the universe. As far as has been observed, there is a striking imbalance of matter over antimatter throughout the visible universe which is irreconcilable with the notion that particles are always created simultaneously with their associated antiparticles. A number of experimentally unverified theories have been posited to attempt to explain this discrepancy.

The Heisenberg uncertainty principle can be formulated in a manner that includes a relationship between time and energy that is relevant to the creation of particle-antiparticle pairs.

$$\Delta E \, \Delta t = \frac{\hbar}{2}$$

It is theorized that this form of the uncertainty principle indicates that a particle and antiparticle pair of energy ΔE can be created spontaneously, so long as its existence is limited to the time Δt. These so-called virtual particles can be created and annihilated even in a vacuum.

COMPETENCY 23.0 UNDERSTAND THE PRINCIPLES OF RADIOACTIVITY AND NUCLEAR REACTIONS AND THEIR APPLICATIONS

Skill 23.1 Applying principles of the conservation of mass number and charge to balance equations for nuclear reactions

In alpha decay, an atom emits an alpha particle. An alpha particle contains two protons and two neutrons. This makes it identical to a helium nucleus and so an alpha particle may be written as He^{2+} or it can be denoted using the Greek letter α. Because a nucleus decaying through alpha radiation loses protons and neutrons, the mass of the atom loses about 4 Daltons and it actually becomes a different element (transmutation). For instance:

$$^{238}U \rightarrow {}^{234}Th + \alpha$$

This isotope of uranium weighs 238 Daltons. Because it is uranium, it has 92 protons, meaning it must have 146 neutrons. When it undergoes alpha decay it loses 2 protons and 2 neutrons. The alpha particle weighs 4 Daltons and the nucleus that has undergone decay weighs 234 Daltons. Thus mass is conserved. The decayed nucleus will have a charge reduced by that of 2 protons following the decay. However, the emitted alpha particle carries the additional charge due to its 2 protons. Thus both charge and mass are conserved over all.

Like alpha decay, beta decay involves emission of a particle. In this case, though, it is a beta particle, which is either an electron or positron. Note that a positron is the antimatter equivalent of an electron and so these particles are often denoted β^- and β^+. Beta plus and minus decay occur via roughly opposite paths. In beta minus decay, a neutron is converted in a proton (specifically, a down quark is converted to an up quark), an electron and an anti-neutrino; the latter two are emitted. In beta plus decay, on the other hand, a proton is converted to a neutron, a positron, and a neutrino; again, the latter two are emitted. As in alpha decay, a nucleus undergoing beta decay is transmuted into a different element because the number of protons is altered. However, because the total number of nucleons remains unchanged, the atomic mass remains the same (note, that the neutron is actually slightly heavier than a proton so mass is gained during beta plus decay).

An examples of beta minus decay is as follows:

$$^{137}_{55}Cs \rightarrow ^{137}_{56}Ba + e^- + \bar{v}_e \quad \text{(beta minus decay)}$$

This isotope of caesium weighs 137 Daltons and, like all caesium isotopes, it has 55 protons. When it undergoes beta minus decay, a neutron is converted to a proton and an electron and an anti-neutrino are lost. The total mass-energy of the system is conserved since the difference in mass between a neutron and an electron plus proton is balanced by the energy of the emitted electron and the anti-neutrino. In beta minus decay a neutron, with no charge, is split into a positively charged proton and a negatively charged electron. Thus the conservation of charge is satisfied. The electron is emitted, while the proton remains in the nucleus. With one extra proton, the nucleus is now a barium isotope.

In order to conserve mass-energy of the system, beta plus decay cannot occur in isolation but only in a nucleus since the mass of a neutron is greater than the mass of a proton plus electron. The difference in binding energy of the mother and daughter nucleus provides the additional energy needed for the reaction to go through. In the example below, charge is conserved when a positively charged proton is converted into a positively charged positron. With one fewer proton, the decayed nucleus is now a neon isotope weighing 22 Daltons.

$$^{22}_{11}Na \rightarrow ^{22}_{10}Ne + e^+ + v_e \quad \text{(beta plus decay)}$$

Problems involving balancing nuclear equations can involve simple radioactive decay, fission, fusion, and other nuclear processes. In all cases, both mass-energy and charge must be balanced.

Problem: Uranium 235 is used as a nuclear fuel in a chain reaction. The reaction is initiated by a single neutron and produces barium 141, an unknown isotope, and 3 neutrons that can go on to propagate the chain reaction. Determine the unknown isotope. Assume that the kinetic energies and the energy released in the reaction is negligible compared to the masses of the isotopes produced.

Solution: We can begin by writing out the reaction, leaving open the unknown isotope X.

$$^{235}_{92}U + ^{1}_{0}n \rightarrow ^{141}_{56}Ba + X + 3^{1}_{0}n$$

We begin with the charge balance; since the neutron has no charge, the unknown isotope must have 36 protons. Consulting a periodic table, we see that this will mean it is a Krypton isotope. Now we can balance the mass. Since the original nucleus weighed 235 Daltons and one neutron was added to it, the total mass of the resultant nuclei must be 236. So, we can simple subtract the weight of the barium isotope and the 3 new neutrons to find the unknown weight:

$$236-141-3=92$$

Thus our unknown isotope is krypton 92, making the balanced equation:

$$^{235}_{92}U + ^{1}_{0}n \rightarrow ^{141}_{56}Ba + ^{92}_{36}Kr + 3^{1}_{0}n$$

Skill 23.2 Analyzing radioactive decay including the half-life concept and the nuclear disintegration series for a given isotope

While the radioactive decay of an individual atom is impossible to predict, a mass of radioactive material will decay at a specific rate. Radioactive isotopes exhibit exponential decay and we can express this decay in a useful equation as follows:

$$A = A_0 e^{kt}$$

Where A is the amount of radioactive material remaining after time t, A_0 is the original amount of radioactive material, t is the elapsed time, and k is the unique activity of the radioactive material. Note that k is unique to each radioactive isotope and it specifies how quickly the material decays. Sometimes it is convenient to express the rate of decay as half-life. A half-life is the time needed for half a given mass of radioactive material to decay. Thus, after one half-life, 50% of an original mass will have decayed, after two half lives, 75% will have decayed and so on.

Let's examine a sample problem related to radioactive decay.

Problem: Radiocarbon dating has been used extensively to determine the age of fossilized organic remains. It is based on the fact that while most of the carbon atoms in living things is ^{12}C, a small percentage is ^{14}C. Since ^{14}C is a radioactive isotope, it is lost from a fossilized specimen at a specific rate following the death of an organism. The original and current mass of ^{14}C can be inferred from the relative amount of ^{12}C. So, if the half-life of ^{14}C is 5730 years and a specimen that originally contained 1.28 mg of ^{14}C now contains 0.10 mg, how old is the specimen?

In certain problems, we may be simply provided with the activity, k, but in this problem we must use the information given about half-life to solve for k.

Since we know that after one half-life, 50% of the material remains radioactive, we can plug into the governing equation above:

$$A = A_0 e^{kt}$$

$$0.5 A_0 = A_0 e^{5730k}$$

$$k = (\ln(0.5))/5730 = -0.0001209$$

Having determined k, we can use this same equation again to determine how old the specimen described above must be:

$$A = A_0 e^{kt}$$

$$0.10 = 1.28 e^{-0.0001209t}$$

$$t = \frac{\ln\left(\frac{0.10}{1.28}\right)}{-0.0001209} = 21087$$

Thus, the specimen is 21,087 years old.

Note that this same equation can be used to calculate the half-life of an isotope if information regarding the decay after a given number of years were provided.

A radioactive isotope often decays into another element that is radioactive as well and continues to decay into a third element. The process continues until a stable element is reached. This chain of disintegration is known as a **decay chain** or a **nuclear disintegration series**. For example, Uranium 238 decays into Radium 226 which decays further into Radon 222.

Skill 23.3   Describing the basic operation of types of radiation detectors

Radiation detectors have been designed for a number of different applications and rely on various mechanisms for converting radiation into an electrical signal. Perhaps the most well-known form of radiation detector is the **Geiger counter**, a device based on the Geiger-Müller (GM) tube. The GM tube is an isolated chamber, usually filled with a combination of an inert gas, such as helium or neon, and a halogen gas, such as chlorine. Inside the tube, an anode and a cathode are present with a strong electric potential difference of hundreds of volts. When ionizing radiation enters the tube, some of the gas molecules are ionized with the positive ion being attracted towards the negatively charged anode and the negative ion (an electron) being attracted towards the positively charged cathode. The strong electric potential causes the ions to gain sufficient kinetic energy to further ionize other gas particles resulting in a cascade of particles and a detectable current. Depending on the specific materials used for the Geiger counter (e.g., the detection "window"), the device can detect alpha particles (helium nuclei), beta particles (an electron or positron resulting from a nuclear beta decay) or gamma rays (high-energy photons). So-called proportional counters, which operate on similar principles, are able to determine the energy of a particle of ionizing radiation through counting the "avalanches" that occur when gas molecules in the tube are ionized and accelerated to cause further ionization.

Photomultiplier tubes are an often-used method for detection of electromagnetic radiation in the visible region of the spectrum as well as in portions of the infrared and ultraviolet regions. The photoelectric effect is the driving phenomenon behind the operation of these devices. Photons that strike a so-called photocathode can cause the emission of an electron which is then accelerated towards a positively charged electrode called a dynode. A number of progressively more positively charged dynodes are arranged in the photomultiplier tube to cause a cascading effect. Each dynode produces further electrons in response to the impact of the incident electron, then "focuses" the emitted electrons onto the next dynode. Charge is collected at an anode, resulting in an electrical signal. The signal current produced by a photomultiplier tube is proportional to the intensity of the incident light.

Another device for detecting ionizing radiation is the **scintillation counter**. In this case a so-called scintillator is used. This is a material that produces a burst of light as a result of incident high-energy particles (such as photons). The incident high-energy particle loses some of its energy in the scintillator and the energy is converted into a number of lower-energy photons. A photomultiplier tube is then used to measure the intensity of the light pulse which yields a measure of the energy of the incident particle. Scintillation counters are used in such contexts as high-energy physics for detection equipment in particle accelerators.

Another radiation detector used in particle physics is the **calorimeter (or microcalorimeter)**. High-energy particles, rather than producing light, as with the scintillator, lose some of their energy to produce heat through incidence on the material used in the calorimeter. A thermometer is used to detect the change in the temperature of the material with the signal intensity providing information on the energy of the particle.

A number of other devices can be used in various contexts involving detection of radiation. Particle track detectors, such as bubble chambers or other devices containing solid, liquid or gaseous materials can be used to identify the passage of charged (or sometimes uncharged) particles. Trails left in the material due to incidence of various particles, which are influenced by a magnetic field, can be analyzed to determine the energy, charge and other features of the particles.

Skill 23.4 Analyzing characteristics of fission and components of a nuclear reactor (e.g., moderator, fuel rods, control rods), including the problems associated with operating a nuclear reactor

Nuclear fission involves the splitting of an atomic nucleus with a subsequent emission of energy in the form of photons with a specific frequency or other particles with a specific kinetic energy. As with combustion, the energy released from a fission reaction can be used to power turbines and to generate electricity. About 20% of the electrical power generated in the United States comes from nuclear reactors.

If controlled, the neutron flux resulting from spontaneously decaying radioactive materials can be used to instigate further decay. With Uranium (^{235}U, specifically), for example, a neutron from another decay can cause the atom to split, thus releasing several neutrons in addition to energy. These neutrons can also cause further decay, resulting in a **chain reaction**. Nuclear weapons involve an uncontrolled chain reaction that yields a tremendously powerful explosion; nuclear power generators take advantage of the same effect, but in a controlled manner.

In order to produce a chain reaction, rods filled with **nuclear fuel pellets** (such as ^{235}U) are used inside the reactor core. Since fission requires relatively slow-moving neutrons, a so-called **moderator** is needed to slow down the fast neutrons produced from fission elsewhere in the core, thus making the chain reaction possible. The moderator can be in the form of water (standard water or deuterium-based "heavy water"), which can also serve as a coolant. While these steps are necessary to sustain a nuclear reaction, it is also equally important to control the reaction. This is accomplished through the use of **control rods** which are filled with a neutron-absorbing material such as boron or silver. While these materials capture free neutrons, they are not themselves fissile and thus their presence limits the nuclear reaction in the core. The location of the control rods can be adjusted for dynamic control of the reaction. The entire apparatus of fuel rods, control rods and moderator (and coolant) is contained inside a pressure vessel.

The heat produced from the controlled nuclear reaction can be used in the same manner as heat from combustion to produce steam and turn a turbine, thus generating electricity. The hyperboloid towers typically seen at nuclear power plants serve the purpose of cooling.

While nuclear power generation does not produce the same kind of carbon-based emissions as combustion of oil or coal, for example, there are other equally (if not more so) undesirable by-products. Waste products from the nuclear reaction are, themselves, often highly radioactive, posing an environmental threat. Ionizing radiation from such waste can cause tremendous damage to living organisms. Since there is no apparent way to easily neutralize these by-products, they must be stored securely to prevent contamination of the environment. This, of course, requires a storage facility and adequate isolation. Security and structural integrity of these facilities must be continually monitored.

Another issue revolving around the use of nuclear power is the ability to use a nuclear reactor to produce weapons-grade materials for use in explosive nuclear devices. This poses a tremendous threat since nuclear weapons are able to cause untold destruction as seen at the close of the Second World War in Japan. Furthermore, in the context of international terrorism, nuclear power plants in non-aggressive nations can still be a threat if attacked. Thus, the plants themselves, although they may be operating for only peaceful purposes, could be damaged in such a way that nuclear materials are released into the environment, or, in a worst-case scenario, a runaway nuclear reaction is initiated. Even otherwise benign human error can lead to catastrophic consequences. Thus, in light of these considerations, a debate continues concerning nuclear power as the dangers and the benefits are weighed.

Skill 23.5 Analyzing fusion reactions and their application to the sun's energy and astronomy

Fusion and fission are two complementary reactions with fusion being the preferred reaction for elements lighter than iron (atomic number 26) or nickel (atomic number 28) and fission being the preferred reaction for elements heavier than iron or nickel. In the case of fusion, energy is released for the lighter elements but is absorbed for the heavier elements. Fission operates in the reverse manner.

Fusion reactions and binding energy (mass defect)

The energy released during a fusion reaction is the binding energy which corresponds to a conversion of some of the mass in the constituent particles into emitted energy (mass defect). The mass defect allows the fused nucleus to exist at a lower energy level, thus making it more stable. Although the energy required to fuse two light nuclei is considerable, the energy released by the reaction is even greater and, under the right circumstances, allows for a self-sustaining (chain) reaction. Such a self-sustaining reaction requires that the energy needed to bring two nuclei to within sufficient proximity for fusion to take place is less than the energy liberated when the nuclei are fused. This initial required energy is necessary to overcome the electrostatic repulsion between the positively charged nuclei.

Fusion is believed to be the primary source of solar energy. In a more terrestrial context, fusion has been exploited for the purpose of creating the so-called hydrogen bomb, which involves an uncontrolled chain reaction that releases an immense amount of destructive energy. In the context of the Sun, fusion provides the energy that is a major factor in maintaining the habitability of the Earth.

Stellar fusion

Since fusion takes place in stars, it is a process that is central to the study of astronomy. Two important reactions involved in stellar fusion are the so-called proton-proton (PP) chain and carbon-nitrogen-oxygen (CNO) cycle. The PP chain involves a process whereby protons are converted into heavier helium nuclei with the concomitant release of energy. It is the primary engine of energy production in stars similar to the Sun. The CNO cycle involves interconversion of carbon, nitrogen and oxygen nuclei and is the driving process in larger stars. Although the PP fusion process takes place through a number of stages, the basic components and products of the reaction are described below, where protons (^{1}H) combine to form a helium nucleus (^{4}He), two neutrinos (v), two positrons (e^+) and energy in the form of photons.

$$4\ ^1H \rightarrow\ ^4He + 2v + 2e^+ + \text{Energy}$$

Skill 23.6 Applying the principle of conservation of mass-energy to calculate nuclear mass defect and binding energy

Independently of one another, mass and energy are not necessarily conserved, as one may be converted to another in certain instances. Together, however, as mass-energy, they are indeed conserved. In most macroscopic circumstances, mass and energy are each conserved individually. On a microscopic scale, however, especially in the realm of particle physics, interconversion of mass and energy can take place to a significant extent. In these cases, the total conservation of mass-energy in a closed system is the most general conservation law.

Binding energy and nuclear mass defect

The binding energy and nuclear mass defect are two complementary concepts associated with the interaction of fundamental particles. The protons that form the nucleus of an atom, for example, are positively charged and, thus, have a mutual electrostatic repulsion. As a result, the energy of a system of closely packed protons is high. Nevertheless, the atomic nucleus is still quite stable. This stability results from a release of energy upon formation of the nucleus, thus lowering the total energy of the system. The released energy is called the binding energy. Conversely, for the reverse process, the binding energy is the energy required to split the nucleus into its component particles, and is often described generally as an energy per nucleon.

The source of the released binding energy during a fusion of protons (or protons and neutrons) into a nucleus is a portion of the mass of the system. According to special relativity, the mass of an object (whether rest or relativistic mass) has an equivalent energy ($E = mc^2$ in the case of rest mass). In the context of nuclear physics, protons that are fused into a nucleus are able to release energy at the expense of a portion of their mass. This nuclear mass defect is the emission of a certain portion of the mass of the constituent particles, in the form of energy, to lower the total energy of the whole (e.g., the nucleus). Quantum chromodynamics (QCD) provides a more fundamental explanation of this phenomenon through the strong nuclear force, or color force. QCD details the interactions of the quarks that compose the substructure of protons and neutrons by way of the color charge, which is a property of quarks, and gluons, which are the mediating boson for the strong force.

Conservation of mass-energy

In light of these basic concepts surrounding the conservation of mass-energy, the nuclear mass defect and binding energy can be calculated using the relativistic relationship of mass and energy. The calculation can be performed by noting the difference in mass between the constituent particles and the final product of a reaction.

Problem: What is the binding energy of a deuteron (a nucleus composed of a proton and a neutron)?

Solution: The binding energy is the equivalent energy resulting from the difference in mass between the deuteron and the constituent particles.

Proton mass = 1.6726×10^{-27} kg
Neutron mass = 1.6749×10^{-27} kg
Deuteron mass = 3.3436×10^{-27} kg

Mass difference: $\Delta m = (1.6726 \times 10^{-27} + 1.6749 \times 10^{-27})\text{kg} - 3.3436 \times 10^{-27}\text{kg}$
$\Delta m = 3.9 \times 10^{-30}$ kg

Binding energy: $\Delta mc^2 = 3.9 \times 10^{-30}\text{kg} \cdot (2.9979 \times 10^8 \text{ m/s})^2 = 3.5051 \times 10^{-13}$ J

Sample Test

DIRECTIONS: Read each item and select the best response.

1. Which of the following theories can be best described as the result of a scientific revolution in physics in the early decades of the twentieth century?
   (Easy) (Skill 1.1 and 1.3)

   A. String theory

   B. Classical electrodynamics

   C. Quantum theory

   D. Galilean relativity

2. Which phenomenon was first explained using the concept of quantization of energy, thus providing one of the key foundational principles for the later development of quantum theory?
   (Rigorous) (Skill 1.2)

   A. The photoelectric effect

   B. Time dilation

   C. Blackbody radiation

   D. Magnetism

3. Which of the following theories may be said to have had the largest technological impact on human culture?
   (Average Rigor) (Skills 1.4 and 1.7)

   A. The Standard Model

   B. General relativity

   C. Quantum mechanics

   D. String theory

4. Which of the following aspects of the use of computers for collecting experimental data is not a concern for the scientist?
   (Rigorous) (Skills 1.7 and 1.8)

   A. The relative speeds of the processor, peripheral, memory storage unit and any other components included in data acquisition equipment

   B. The financial cost of the equipment, utilities and maintenance

   C. Numerical error due to a lack of infinite precision in digital equipment

   D. The order of complexity of data analysis algorithms

5. Which description best describes the role of a scientific model of a physical phenomenon?
   (Average Rigor) (Skill 2.1)

   A. An explanation that provides a reasonably accurate approximation of the phenomenon
   B. A theoretical explanation that describes exactly what is taking place
   C. A purely mathematical formulation of the phenomenon
   D. A predictive tool that has no interest in what is actually occurring

6. If a particular experimental observation contradicts a theory, what is the most appropriate approach that a physicist should take?
   (Average Rigor) (Skill 2.1)

   A. Immediately reject the theory and begin developing a new theory that better fits the observed results
   B. Report the experimental result in the literature without further ado
   C. Repeat the observations and check the experimental apparatus for any potential faulty components or human error, and then compare the results once more with the theory
   D. Immediately reject the observation as in error due to its conflict with theory

7. Which situation calls might best be described as involving an ethical dilemma for a scientist?
   (Rigorous) (Skill 2.2)

   A. Submission to a peer-review journal of a paper that refutes an established theory
   B. Synthesis of a new radioactive isotope of an element
   C. Use of a computer for modeling a newly-constructed nuclear reactor
   D. Use of a pen-and-paper approach to a difficult problem

8. Which statement best describes a valid approach to testing a scientific hypothesis?
   (Easy) (Skill 2.3)

   A. Use computer simulations to verify the hypothesis
   B. Perform a mathematical analysis of the hypothesis
   C. Design experiments to test the hypothesis
   D. All of the above

9. Which of the following is not a key purpose for the use of open communication about and peer-review of the results of scientific investigations? *(Average Rigor) (Skill 2.4)*

   A. Testing, by other scientists, of the results of an investigation for the purpose of refuting any evidence contrary to an established theory

   B. Testing, by other scientists, of the results of an investigation for the purpose of finding or eliminating any errors in reasoning or measurement

   C. Maintaining an open, public process to better promote honesty and integrity in science

   D. Provide a forum to help promote progress through mutual sharing and review of the results of investigations

10. Which of the following devices would be best suited for an experiment designed to measure alpha particle emissions from a sample? *(Average Rigor) (Skill 3.1)*

    A. Photomultiplier tube

    B. Thermocouple

    C. Geiger-Müller tube

    D. Transistor

11. Which statement best characterizes the relationship of mathematics and experimentation in physics? *(Easy) (Skill 3.2)*

    A. Experimentation has no bearing on the mathematical models that are developed for physical phenomena

    B. Mathematics is a tool that assists in the development of models for various physical phenomena as they are studied experimentally, with observations of the phenomena being a test of the validity of the mathematical model

    C. Mathematics is used to test the validity of experimental apparatus for physical measurements

    D. Mathematics is an abstract field with no relationship to concrete experimentation

12. What best describes an appropriate view of scale in scientific investigation? *(Rigorous) (Skill 3.5)*

   A. Scale is irrelevant: the same fundamental principles apply at all scale levels

   B. Scale has no bearing on experimentation, although it may have a part in some theories

   C. Scale has only a practical effect: it may make certain experiments more or less difficult due to limitations on equipment, but it has no fundamental or theoretical impact

   D. Scale is an important consideration both experimentally and theoretically: certain phenomena may have negligible effect at one scale, but may have an overwhelming effect at another scale

13. Which statement best describes a rationale for the use of statistical analysis in characterizing the numerical results of a scientific experiment or investigation? *(Average Rigor) (Skill 3.6)*

   A. Experimental results need to be adjusted, through the use of statistics, to conform to theoretical predictions and computer models

   B. Since experiments are prone to a number of errors and uncertainties, statistical analysis provides a method for characterizing experimental measurements by accounting for or quantifying these undesirable effects

   C. Experiments are not able to provide any useful information, and statistical analysis is needed to impose a theoretical framework on the results

   D. Statistical analysis is needed to relate experimental measurements to computer-simulated values

14. Which of the following is *not* an SI unit?
    (Average Rigor) (Skill 4.1)

    A. Joule

    B. Coulomb

    C. Newton

    D. Erg

15. Which of the following best describes the relationship of precision and accuracy in scientific measurements?
    (Easy) (Skill 4.3)

    A. Accuracy is how well a particular measurement agrees with the value of the actual parameter being measured; precision is how well a particular measurement agrees with the average of other measurements taken for the same value

    B. Precision is how well a particular measurement agrees with the value of the actual parameter being measured; accuracy is how well a particular measurement agrees with the average of other measurements taken for the same value

    C. Accuracy is the same as precision

    D. Accuracy is a measure of numerical error; precision is a measure of human error

16. Which statement best describes how significant figures must be treated when analyzing a collection of measured data?
    (Rigorous) (Skill 4.4)

    A. Report all data and any results calculated using that data with the number of significant figures that corresponds to the least accurate number (i.e., the one with the fewest significant figures)

    B. Report all data and any results calculated using that data with the number of significant figures that corresponds to the most accurate number (i.e., the one with the most significant figures)

    C. Follow appropriate rules for the propagation of error, maintaining an appropriate number of significant figures for each particular set of data or calculated result

    D. Report all data with the appropriate number of significant figures, but ignore error when calculating results

17. Which of the following experiments presents the most likely cause for concern about laboratory safety?
(Average Rigor) (Skill 4.6)

A. Computer simulation of a nuclear reactor

B. Vibration measurement with a laser

C. Measurement of fluorescent light intensity with a battery-powered photodiode circuit

D. Ambient indoor ionizing radiation measurement with a Geiger counter.

18. Which of the following mathematical tools would not typically be used for the analysis of an electromagnetic phenomenon?
(Rigorous) (Skills 5.1-5.5)

A. Trigonometry

B. Vector calculus

C. Group theory

D. Numerical methods

19. For a problem that involves parameters that vary in rate with direction and location, which of the following mathematical tools would most likely be of greatest value?
(Rigorous) (Skill 5.5)

A. Trigonometry
B. Numerical analysis
C. Group theory
D. Vector calculus

20. When acceleration is plotted versus time, the area under the graph represents:
(Average Rigor) (Skill 6.3)

A. Time
B. Distance
C. Velocity
D. Acceleration

21. A brick and hammer fall from a ledge at the same time. They would be expected to:
(Easy) (Skill 6.4)

A. Reach the ground at the same time

B. Accelerate at different rates due to difference in weight

C. Accelerate at different rates due to difference in potential energy

D. Accelerate at different rates due to difference in kinetic energy

22. Given a vase full of water, with holes punched at various heights. The water squirts out of the holes, achieving different distances before hitting the ground. Which of the following accurately describes the situation?
(Average Rigor) (Skill 7.6)

   A. Water from higher holes goes farther, due to Pascal's Principle.

   B. Water from higher holes goes farther, due to Bernoulli's Principle.

   C. Water from lower holes goes farther, due to Pascal's Principle.

   D. Water from lower holes goes farther, due to Bernoulli's Principle.

23. The magnitude of a force is:
(Easy) (Skill 8.1)

   A. Directly proportional to mass and inversely to acceleration

   B. Inversely proportional to mass and directly to acceleration

   C. Directly proportional to both mass and acceleration

   D. Inversely proportional to both mass and acceleration

24. If a force of magnitude F gives a mass M an acceleration A, then a force $3F$ would give a mass $3M$ an acceleration:
(Average Rigor) (Skill 8.2)

   A. A

   B. $12A$

   C. $A/2$

   D. $6A$

25. An inclined plane is tilted by gradually increasing the angle of elevation θ, until the block will slide down at a constant velocity. The coefficient of friction, μ_k, is given by:
(Rigorous) (Skill 8.3)

   A. cos θ

   B. sin θ

   C. cosecant θ

   D. tangent θ

26. A mass of 2 kg connected to a spring undergoes simple harmonic motion at a frequency of 3 Hz. What is the spring constant?
(Average Rigor) (Skill 8.4)

   A. 6 kg/s^2

   B. 18 kg/s^2

   C. 710 kg/s^2

   D. 1000 kg/s^2

27. A projectile with a mass of 1.0 kg has a muzzle velocity of 1500.0 m/s when it is fired from a cannon with a mass of 500.0 kg. If the cannon slides on a frictionless track, it will recoil with a velocity of ____ m/s.
(Rigorous) (Skill 8.5)

A. 2.4

3.0

3.5

1500

28. If an object of length 5 meters (along the $\hat{y}$ axis) is traveling at 99% of the speed of light in the $\hat{x}$ direction, what is its length, as measured by an observer at the origin?
(Rigorous) (Skill 8.6)

A. A. 0 m

B. 0.71 m

C. 3.4 m

D. 5 m

29. An object traveling through air loses part of its energy of motion due to friction. Which statement best describes what has happened to this energy?
(Easy) (Skill 9.1)

A. The energy is destroyed

B. The energy is converted to static charge

C. The energy is radiated as electromagnetic waves

D. The energy is lost to heating of the air

30. A 100 g mass revolving around a fixed point, on the end of a 0.5 meter string, circles once every 0.25 seconds. What is the magnitude of the centripetal acceleration?
(Average Rigor) (Skill 10.2)

A. A. 1.23 m/s^2

B. 31.6 m/s^2

C. 100 m/s^2

D. 316 m/s^2

31. A satellite is in a circular orbit above the earth. Which statement is false?
Average Rigor) (Skill 10.3)

A. An external force causes the satellite to maintain orbit.

B. The satellite's inertia causes it to maintain orbit.

C. The satellite is accelerating toward the earth.

D. The satellite's velocity and acceleration are not in the same direction.

32. A simple pendulum with a period of one second has its mass doubled. If the length of the string is quadrupled, the new period will be:
    (Rigorous) (Skill 10.4)

    A. 1 second

    B. 2 seconds

    C. 3 seconds

    D. 5 seconds

33. A uniform pole weighing 100 grams, that is one meter in length, is supported by a pivot at 40 centimeters from the left end. In order to maintain static position, a 200 gram mass must be placed _____ centimeters from the left end.
    (Rigorous) (Skill 10.6)

    A. 10

    B. 45

    C. 35

    D. 50

34. Which of the following units is not used to measure torque?
    Average Rigor) (Skill 10.7)

    A. slug ft

    B. lb ft

    C. N m

    D. dyne cm

35. A gas maintained at a constant pressure has a specific heat which is greater than its specific heat when maintained at constant volume, because:
    (Rigorous) (Skill 11.1)

    A. The Coefficient of Expansion changes.

    B. The gas enlarges as a whole.

    C. Brownian motion causes random activity.

    D. Work is done to expand the gas.

36. The First Law of Thermodynamics takes the form dU = dW when the conditions are:
    (Rigorous) (Skill 11.2)

    A. Isobaric

    B. Isochloremic

    C. Isothermal

    D. Adiabatic

37. A temperature change of 40 degrees Celsius is equal to a change in Fahrenheit degrees of:
    (Average Rigor) (Skill 11.4)

    A. 40

    B. 20

    C. 72

    D. 112

PHYSICS

38. A cooking thermometer in an oven works because the metals it is composed of have different:
(Average Rigor) (Skill 11.5)

A. Melting points

B. Heat convection

C. Magnetic fields

D. Coefficients of expansion

39. Heat transfer by electromagnetic waves is termed:
(Easy) (Skill 11.6)

A. Conduction

B. Convection

C. Radiation

D. Phase Change

40. A long copper bar has a temperature of 60°C at one end and 0°C at the other. The bar reaches thermal equilibrium (barring outside influences) by the process of heat:
(Average Rigor) (Skill 11.6)

A. Fusion

B. Convection

C. Conduction

D. Microwaving

41. Use the information on heats below to solve this problem. An ice block at 0° Celsius is dropped into 100 g of liquid water at 18° Celsius. When thermal equilibrium is achieved, only liquid water at 0° Celsius is left. What was the mass, in grams, of the original block of ice?

Given: Heat of fusion of ice = 80 cal/g
Heat of vaporization of ice = 540 cal/g
Specific Heat of ice = 0.50 cal/g°C
Specific Heat of water = 1 cal/g°C

(Rigorous) (Skill 11.7)

A. 2.0

B. 5.0

C. 10.0

D. 22.5

42. Which of the following is not an assumption upon which the kinetic-molecular theory of gases is based?
(Rigorous) (Skill 11.8)

A. Quantum mechanical effects may be neglected

B. The particles of a gas may be treated statistically

C. The particles of the gas are treated as very small masses

D. Collisions between gas particles and container walls are inelastic

43. What should be the behavior of an electroscope, which has been grounded in the presence of a positively charged object (1), after the ground connection is removed and then the charged object is removed from the vicinity (2)?
(Average Rigor) (Skill 12.1)

A. The metal leaf will start deflected (1) and then relax to an undeflected position (2)

B. The metal leaf will start in an undeflected position (1) and then be deflected (2)

C. The metal leaf will remain undeflected in both cases

D. The metal leaf will be deflected in both cases

44. Static electricity generation occurs by:
(Easy) (Skill 12.2)

A. Telepathy

B. Friction

C. Removal of heat

D. Evaporation

45. The electric force in Newtons, on two small objects (each charged to –10 microCoulombs and separated by 2 meters) is:
(Rigorous) (Skill 12.6)

A. 1.0

B. 9.81

C. 31.0

D. 0.225

46. When the current flowing through a fixed resistance is doubled, the amount of heat generated is:
(Average Rigor) (Skill 13.1 and 13.4)

A. Quadrupled

B. Doubled

C. Multiplied by pi

D. Halved

47. A semi-conductor allows current to flow:
(Easy) (Skill 13.3)

A. Never

B. Always

C. As long as it stays below a maximum temperature

D. When a minimum voltage is applied

48. A 10 ohm resistor and a 50 ohm resistor are connected in parallel. If the current in the 10 ohm resistor is 5 amperes, the current (in amperes) running through the 50 ohm resistor is:
(Rigorous) (Skills 13.1 and 13.4)

A. 1

B. 50

C. 25

D. 60

49. How much power is dissipated through the following resistive circuit?
(Average Rigor) (Skill 13.6 and 13.7)

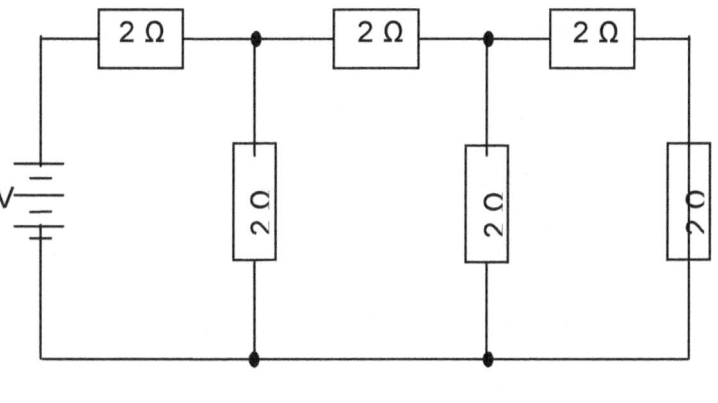

A. 0 W

B. 0.22 W

C. 0.31 W

D. 0.49 W

50. The greatest number of 100 watt lamps that can be connected in parallel with a 120 volt system without blowing a 5 amp fuse is:
(Rigorous) (Skill 13.7)

A. 24

B. 12

C. 6

D. 1

51. What is the electric flux density (or electric displacement) through each face of a cube, of side length 2 meters, that contains a central point charge of 2 Coulombs?
(Rigorous) (Skill 13.8)

A. 0.50 Coulomb/m^2

B. 0.33 Coulomb/m^2

C. 0.13 Coulomb/m^2

D. 0.33 Tesla

52. All of the following use semi-conductor technology, except a(n):
(Average Rigor) (Skill 13.9)

A. Transistor

B. Diode

C. Capacitor

D. Operational Amplifier

53. What is the direction of the magnetic field at the center of the loop of current (I) shown below (i.e., at point A)?
(Easy) (Skill 14.1)

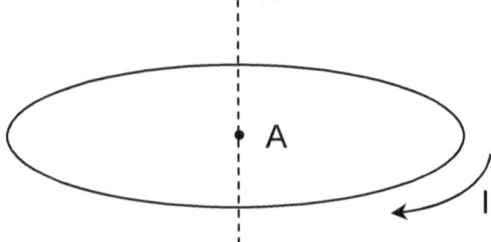

A. Down, along the axis (dotted line)
B. Up, along the axis (dotted line)
C. The magnetic field is oriented in a radial direction
D. There is no magnetic field at point A

54. What effect might an applied external magnetic field have on the magnetic domains of a ferromagnetic material?
(Rigorous) (Skill 14.2)

A. The domains that are not aligned with the external field increase in size, but those that are aligned decrease in size
B. The domains that are not aligned with the external field decrease in size, but those that are aligned increase in size
C. The domains align perpendicular to the external field
D. There is no effect on the magnetic domains

55. Which of the following statements may be taken as a legitimate inference based upon the Maxwell equation that states $\nabla \cdot \mathbf{B} = 0$?
(Average Rigor) (Skill 14.6)

A. The electric and magnetic fields are decoupled
B. The electric and magnetic fields are mediated by the W boson
C. There are no photons
D. There are no magnetic monopoles

56. The use of two circuits next to each other, with a change in current in the primary circuit, demonstrates:
(Rigorous) (Skill 15.1)

A. Mutual current induction
B. Dielectric constancy
C. Harmonic resonance
D. Resistance variation

57. A static magnetic flux density of 1 Tesla is linked by a wire loop, as shown below. What is the electromotive force (EMF) around the loop?
(Average Rigor) (Skill 15.1)

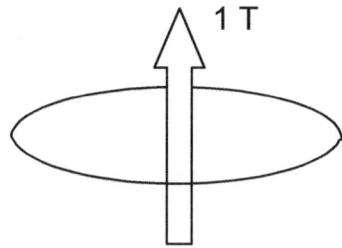

A. 1 Volt

B. 2 Volts

C. 5 Volts

D. 0 Volts

58. What is the effect of running current in the same direction along two parallel wires, as shown below?
(Rigorous) (Skill 15.2)

A. There is no effect

B. The wires attract one another

C. The wires repel one another

D. A torque is applied to both wires

59. A wave generator is used to create a succession of waves. The rate of wave generation is one every 0.33 seconds. The period of these waves is:
(Average Rigor) (Skills 16.1 and 16.3)

A. 2.0 seconds

B. 1.0 seconds

C. 0.33 seconds

D. 3.0 seconds

60. An electromagnetic wave propagates through a vacuum. Independent of its wavelength, it will move with constant:
(Easy) (Skill 16.2)

A. Acceleration

B. Velocity

C. Induction

D. Sound

61. A wave has speed 60 m/s and wavelength 30,000 m. What is the frequency of the wave?
(Average Rigor) (Skill 16.3)

A. 2.0×10^{-3} Hz

B. 60 Hz

C. 5.0×10^2 Hz

D. 1.8×10^6 Hz

62. The wave phenomenon of polarization applies only to:
(Average Rigor) (Skill 16.5)

   A. Longitudinal waves

   B. Transverse waves

   C. Sound

   D. Light

63. If one sound is ten decibels louder than another, the ratio of the intensity of the first to the second is:
(Average Rigor) (Skill 17.1)

   A. 20:1

   B. 10:1

   C. 1:1

   D. 1:10

64. The velocity of sound is greatest in:
(Average Rigor) (Skill 17.2)

   A. Water

   B. Steel

   C. Alcohol

   D. Air

65. Which of the following apparatus can be used to measure the wavelength of a sound produced by a tuning fork?
(Average Rigor) (Skill 17.3)

   A. A glass cylinder, some water, and iron filings

   B. A glass cylinder, a meter stick, and some water

   C. A metronome and some ice water

   D. A comb and some tissue

66. A vibrating string's frequency is _____ proportional to the _____.
(Rigorous) (Skill 17.4)

   A. Directly; Square root of the tension

   B. Inversely; Length of the string

   C. Inversely; Squared length of the string

   D. Inversely; Force of the plectrum

67. The highest energy is associated with:
(Easy) (Skill 18.1)

   A. UV radiation

   B. Yellow light

   C. Infrared radiation

   D. Gamma radiation

68. In order to switch between two different reference frames in special relativity, we use the _____ transformation.
(Average Rigor) (Skill 18.5)

A. Galilean

B. Lorentz

C. Euclidean

D. Laplace

69. A crew is on-board a spaceship, traveling at 60% of the speed of light with respect to the earth. The crew measures the length of their ship to be 240 meters. When a ground-based crew measures the apparent length of the ship, it equals:
(Rigorous) (Skill 18.5)

A. 400 m

B. 300 m

C. 240 m

D. 192 m

70. All of the following phenomena are considered "refractive effects" except for:
(Rigorous) (Skills 16.5 and 18.6)

A. The red shift

B. Total internal reflection

C. Lens dependent image formation

D. Snell's Law

71. A stationary sound source produces a wave of frequency F. An observer at position A is moving toward the horn, while an observer at position B is moving away from the horn. Which of the following is true?
(Rigorous) (Skill 18.6)

A. $F_A < F < F_B$

B. $F_B < F < F_A$

C. $F < F_A < F_B$

D. $F_B < F_A < F$

72. Automobile mirrors that have a sign, "objects are closer than they appear" say so because:
(Rigorous) (Skill 19.1)

A. The real image of an obstacle, through a converging lens, appears farther away than the object.

B. The real or virtual image of an obstacle, through a converging mirror, appears farther away than the object.

C. The real image of an obstacle, through a diverging lens, appears farther away than the object.

D. The virtual image of an obstacle, through a diverging mirror, appears farther away than the object.

73. An object two meters tall is speeding toward a plane mirror at 10 m/s. What happens to the image as it nears the surface of the mirror?
(Rigorous) (Skill 19.2)

   A. It becomes inverted.

   B. The Doppler Effect must be considered.

   C. It remains two meters tall.

   D. It changes from a real image to a virtual image.

74. If an object is 20 cm from a convex lens whose focal length is 10 cm, the image is:
(Rigorous) (Skill 19.3)

   A. Virtual and upright

   B. Real and inverted

   C. Larger than the object

   D. Smaller than the object

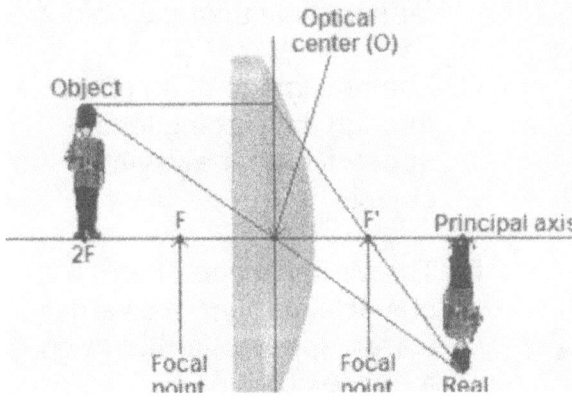

75. A monochromatic ray of light passes from air to a thick slab of glass (n=1.41) at an angle of 45° from the normal. At what angle does it leave the air/glass interface?
(Rigorous) (Skill 19.4)

   A. 45°

   B. 30°

   C. 15°

   D. 55°

76. A quantum of light energy is called a:
(Easy) (Skill 20.1)

   A. Dalton

   B. Photon

   C. Curie

   D. Heat Packet

77. Which statement best describes why population inversion is necessary for a laser to operate?
(Rigorous) (Skill 20.3)

A. Population inversion prevents too many electrons from being excited into higher energy levels, thus preventing damage to the gain medium.

B. Population inversion maintains a sufficient number of electrons in a higher energy state so as to allow a significant amount of stimulated emission.

C. Population inversion prevents the laser from producing coherent light.

D. Population inversion is not necessary for the operation of most lasers.

78. The constant of proportionality between the energy and the frequency of electromagnetic radiation is known as the:
(Easy) (Skills 20.1 and 20.5)

A. Rydberg constant

B. Energy constant

C. Planck constant

D. Einstein constant

79. The initial position and velocity of an electron (mass=9.1×10^{-31} kg) are measured. If the position is known to within a nuclear radius (10^{-14}m) and the velocity is known with the smallest accuracy allowed by the uncertainty principle, what will its position uncertainly be one year later? (Assume no force acts on the electron and h=6.6×10^{-34} J·s)
(Rigorous) (Skill 20.6)

A. 2.28×10^{18}m

B. 7.27×10^{17}m

C. 1.15×10^{10} m

D. 3.3×10^{17} m

80. Bohr's theory of the atom was the first to quantize:
(Average Rigor) (Skill 21.1)

A. Work

B. Angular Momentum

C. Torque

D. Duality

81. Two neutral isotopes of a chemical element have the same numbers of:
    (Easy) (Skill 21.2)

    A. Electrons and Neutrons

    B. Electrons and Protons

    C. Protons and Neutrons

    D. Electrons, Neutrons, and Protons

82. When an electron is "orbiting" the nucleus in an atom, it is said to possess an intrinsic spin (spin angular momentum). How many values can this spin have in any given electron?
    (Average Rigor) (Skill 22.1)

    A. 1

    B. 2

    C. 3

    D. 8

83. Which of the following statements best describes the diagram below?
    (Average Rigor) (Skill 22.1)

    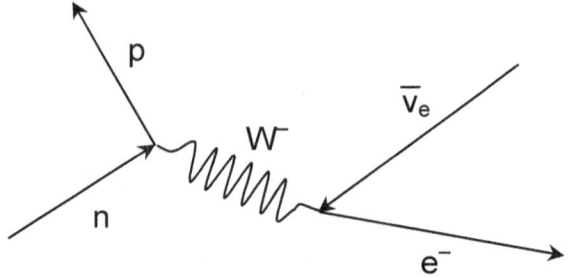

    A. A depiction of the mechanism for gamma ray emission

    B. A Feynman diagram of a photon exchange

    C. A Feynman diagram of a beta decay

    D. A depiction of the classical concept of electromagnetism

84. Electrons are:
    (Easy) (Skill 22.5)

    A. More massive than neutrons

    B. Positively charged

    C. Neutrally charged

    D. Negatively charged

85. When a radioactive material emits an alpha particle only, its atomic number will:
    (Average Rigor) (Skill 23.1)

    A. Decrease

    B. Increase

    C. Remain unchanged

    D. Change randomly

86. Ten grams of a sample of a radioactive material (half-life = 12 days) were stored for 48 days and re-weighed. The new mass of material was:
    (Rigorous) (Skill 23.2)

    A. 1.25 g

    B. 2.5 g

    C. 0.83 g

    D. 0.625 g

87. In a nuclear pile, the control rods are composed of:
    (Average Rigor) (Skill 23.4)

    A. Boron

    B. Einsteinium

    C. Isoptocarpine

    D. Phlogiston

88. In a fission reactor, heavy water:
    (Average Rigor) (Skill 23.4)

    A. Cools off neutrons to control temperature

    B. Moderates fission reactions

    C. Initiates the reaction chain

    D. Dissolves control rods

89. Which of the following pairs of elements are not found to fuse in the centers of stars?
    (Average Rigor) (Skill 23.5)

    A. Oxygen and Helium

    B. Carbon and Hydrogen

    C. Beryllium and Helium

    D. Cobalt and Hydrogen

90. Given the following values for the masses of a proton, a neutron and an alpha particle, what is the nuclear binding energy of an alpha particle?

    Proton mass = 1.6726×10^{-27} kg
    Neutron mass = 1.6749×10^{-27} kg
    Alpha particle mass = 6.6465×10^{-27} kg

    (Rigorous) (Skill 23.6)

    A. 0 J

    B. 7.3417×10^{-27} J

    C. 4 J

    D. 4.3589×10^{-12} J

Free Response Question

A 4 kg sphere, having a radius of 0.25 m and a uniform density, is rotating at f = 3 Hertz. If this sphere is dropped into a 2 liter, thermally isolated container of water at 25° Celsius, what is the final temperature of the water when the sphere (and water) stops rotating?

The rotating sphere will lose mechanical energy due to friction, eventually leading to complete transfer of kinetic energy to heat energy. Applying the first law of thermodynamics, the amount of kinetic energy that is lost to friction must be equal to the amount of heat energy added to the water. It is necessary, then, to first calculate the amount of kinetic energy in the rotating sphere.

The rotational kinetic energy of the sphere can be found by first determining the rotational inertia, I, of the sphere. To do this, use the following definition:

$$I = \int \rho^2 \, dm$$

Here, ρ is the distance from the axis of rotation and dm is the differential element of mass. Since the mass is uniformly distributed throughout the sphere, the density σ is constant. The integral for I is most easily evaluated in spherical coordinates. Thus:

$$I = \int_0^\pi \int_0^{2\pi} \int_0^R \sigma (r\sin\theta)^2 r^2 \sin\theta \, dr \, d\phi \, d\theta$$

Here, ρ has been expressed in terms of the radial distance r (the sphere rotates around a diameter – the identity of the diameter is not relevant to the problem). R is the radius of the sphere. Noting that σ is simply the mass of the sphere divided by the volume of the sphere, the integral may be evaluated as follows:

$$I = \frac{4\,\text{kg}}{\frac{4}{3}\pi R^3} \int_0^R r^4 \, dr \int_0^{2\pi} d\phi \int_0^\pi \sin^3\theta \, d\theta$$

$$I = \frac{4\,\text{kg}}{\frac{4}{3}\pi R^3} \left[\frac{R^5}{5}\right][2\pi]\left[-\cos\theta + \frac{\cos^3\theta}{3}\right]_0^\pi$$

$$I = 4\,\text{kg}\,\frac{3R^2}{10}\left[\frac{4}{3}\right]$$

$$I = 4\,\text{kg}\,\frac{2R^2}{5}$$

$$I = 0.1\,\text{kg}\cdot\text{m}^2$$

The rotational kinetic energy is the following:

$$KE_{rotational} = \frac{1}{2}I\omega^2$$
$$KE_{rotational} = 2\pi^2 \, If^2$$
$$KE_{rotational} = 17.8 \, J$$

This is the energy that is transformed, through friction, to heat energy. The specific heat (c) of water is 4190 J kg^{-1} K^{-1}. The density of water is about 1,000 kg m^{-3}. The mass of 2 liters of water is:

$$m = \rho V = 1000 \, \frac{kg}{m^3} \cdot 2L \cdot \frac{1 m^3}{1000 L} = 2 \, kg$$

The following equation expresses the relationship of the initial temperature (T_i) and final temperature (T_f) of the water with respect to the heat energy added (Q).

$$Q = 17.8 \, J = mc(T_f - T_i)$$

This equation may now be solved, once the initial temperature has been converted to Kelvins:

$$17.8 \, J = 2 \, kg \cdot 4190 \, \frac{J}{kg \cdot K}(T_f - 298 \, K)$$
$$0.002 \, K = (T_f - 298 \, K)$$
$$T_f = 298.002 \, K = 25.002°C$$

Thus, although the change in temperature is slight, it can be seen that the rotational energy of the sphere is transferred to heat energy in the water, resulting in a temperature that is higher by 0.002 Kelvins.

Answer Key

1.	C	19.	D	37.	C	55.	D	73.	C
2.	C	20.	C	38.	D	56.	A	74.	B
3.	C	21.	A	39.	C	57.	D	75.	B
4.	D	22.	D	40.	C	58.	B	76.	B
5.	A	23.	C	41.	D	59.	C	77.	B
6.	C	24.	A	42.	D	60.	B	78.	C
7.	B	25.	D	43.	B	61.	A	79.	D
8.	D	26.	C	44.	B	62.	B	80.	B
9.	A	27.	B	45.	D	63.	B	81.	B
10.	C	28.	D	46.	A	64.	B	82.	B
11.	B	29.	D	47.	D	65.	B	83.	C
12.	D	30.	D	48.	A	66.	A	84.	D
13.	B	31.	B	49.	C	67.	D	85.	A
14.	D	32.	B	50.	C	68.	B	86.	D
15.	A	33.	C	51.	B	69.	D	87.	A
16.	C	34.	A	52.	C	70.	A	88.	B
17.	B	35.	D	53.	A	71.	B	89.	D
18.	C	36.	D	54.	B	72.	D	90.	D

Rigor Analysis Table

Easy 19 %	1,8,11,15,21,23,29,39,44,47,53,60,67,76,78,81,84
Average Rigor 41%	3,5,6,9,10,13,14,17,20,22,24,26,30,31,34,37,38,40,43, 46,49,52,55,57,59,61,62,63,64,65,68,80,82,83,85,87,88
Rigorous 40 %	2,4,7,12,16,18,19,25,27,28,32,33,35,36,41,42,45,48,50, 51,54,56,58,66,69,70,71,72,73,74,75,77,79,86,90

TEACHER CERTIFICATION STUDY GUIDE

Rationales with Sample Questions

1. **Which of the following theories can be best described as the result of a scientific revolution in physics in the early decades of the twentieth century?**
   (Easy) (Skill 1.1 and 1.3)

   A. String theory

   B. Classical electrodynamics

   C. Quantum theory

   D. Galilean relativity

   Answer: C

   The early part of the twentieth century is well-known as the time during which quantum theory was largely developed. This theory provided the foundation for the development of many modern technologies, including lasers, computers and various types of instrumentation. The other three responses were either largely developed prior to (B and D) or later than (A) this time period.

2. **Which phenomenon was first explained using the concept of quantization of energy, thus providing one of the key foundational principles for the later development of quantum theory?**
   (Rigorous) (Skill 1.2)

   A. The photoelectric effect

   B. Time dilation

   C. Blackbody radiation

   D. Magnetism

   Answer: C

   Although the photoelectric effect applied principles of quantization in explaining the behavior of electrons emitted from a metallic surface when the surface is illuminated with electromagnetic radiation, the explanation of the phenomenon of blackbody radiation, provided by Max Planck, was the first major success of the concept of quantized energy. Magnetism may be explained quantum mechanically, but such an explanation was not forthcoming until well after Planck's quantization hypothesis. Time dilation is primarily explained through relativity theory.

3. **Which of the following theories may be said to have had the largest technological impact on human culture?**
   (Average Rigor) (Skills 1.4 and 1.7)

       A. The Standard Model

       B. General relativity

       C. Quantum mechanics

       D. String theory

   Answer: C

   Although general relativity (B) and string theory (D) are, to some extent, "buzz words" with a certain level of popular fame, they have not had any extensive technological impact. The Standard Model (A) also has questionable technological significance for truly extensive technological value. Quantum mechanics (C), however, can be credited as the theoretical foundation for semiconductor technologies (such as computers), lasers and numerous types of instrumentation for chemistry and physics.

4. **Which of the following aspects of the use of computers for collecting experimental data is not a concern for the scientist?**
   (Rigorous) (Skills 1.7 and 1.8)

       A. The relative speeds of the processor, peripheral, memory storage unit and any other components included in data acquisition equipment

       B. The financial cost of the equipment, utilities and maintenance

       C. Numerical error due to a lack of infinite precision in digital equipment

       D. The order of complexity of data analysis algorithms

   Answer: D

   Although answer D might be a concern for later, when actual analysis of the data is undertaken, the collection of data typically does not suffer from this problem. The use of computers does, however, pose problems when, for example, a peripheral collects data at a rate faster than the computer can process it (A), or if the cost of running the equipment or of purchasing the equipment is prohibitive (B). Numerical error is always a concern with any digital data acquisition system, since the data that is collected is never exact.

TEACHER CERTIFICATION STUDY GUIDE

5. **Which description best describes the role of a scientific model of a physical phenomenon?**
   (Average Rigor) (Skill 2.1)

   A. An explanation that provides a reasonably accurate approximation of the phenomenon

   B. A theoretical explanation that describes exactly what is taking place

   C. A purely mathematical formulation of the phenomenon

   D. A predictive tool that has no interest in what is actually occurring

   Answer: A

   A scientific model seeks to provide the most fundamental and accurate description possible for physical phenomena, but, given the fact that natural science takes an *a posteriori* approach, models are always tentative and must be treated with some amount of skepticism. As a result, A is a better answer than B. Answers C and D overly emphasize one or another aspect of a model, rather than a synthesis of a number of aspects (such as a mathematical and predictive aspect).

6. **If a particular experimental observation contradicts a theory, what is the most appropriate approach that a physicist should take?**
   (Average Rigor) (Skill 2.1)

   A. Immediately reject the theory and begin developing a new theory that better fits the observed results

   B. Report the experimental result in the literature without further ado

   C. Repeat the observations and check the experimental apparatus for any potential faulty components or human error, and then compare the results once more with the theory

   D. Immediately reject the observation as in error due to its conflict with theory

Answer: C

When experimental results contradict a reigning physical theory, as they do from time to time, it is almost never appropriate to immediately reject the theory (A) *or* the observational results (D). Also, since this is the case, reporting the result in the literature, without further analysis to provide an adequate explanation of the discrepancy, is unwise and unwarranted. Further testing is appropriate to determine whether the experiment is repeatable and whether any equipment or human errors have occurred. Only after further testing may the physicist begin to analyze the implications of the observational result.

7. **Which situation calls might best be described as involving an ethical dilemma for a scientist?**
   (Rigorous) (Skill 2.2)

    A. Submission to a peer-review journal of a paper that refutes an established theory

    B. Synthesis of a new radioactive isotope of an element

    C. Use of a computer for modeling a newly-constructed nuclear reactor

    D. Use of a pen-and-paper approach to a difficult problem

   Answer: B

   Although answer A may be controversial, it does not involve an inherently ethical dilemma, since there is nothing unethical about presenting new information if it is true or valid. Answer C, likewise, has no necessary ethical dimension, as is the case with D. Synthesis of radioactive material, however, involves an ethical dimension with regard to the potential impact of the new isotope on the health of others and on the environment. The potential usefulness of such an isotope in weapons development is another ethical consideration.

8. **Which statement best describes a valid approach to testing a scientific hypothesis?**
   (Easy) (Skill 2.3)

    A. Use computer simulations to verify the hypothesis

    B. Perform a mathematical analysis of the hypothesis

    C. Design experiments to test the hypothesis

    D. All of the above

   Answer: D

   Each of the answers A, B and C can have a crucial part in testing a scientific hypothesis. Although experiments may hold more weight than mathematical or computer-based analysis, these latter two methods of analysis can be critical, especially when experimental design is highly time consuming or financially costly.

TEACHER CERTIFICATION STUDY GUIDE

9. **Which of the following is not a key purpose for the use of open communication about and peer-review of the results of scientific investigations?**
   (Average Rigor) (Skill 2.4)

   A. Testing, by other scientists, of the results of an investigation for the purpose of refuting any evidence contrary to an established theory

   B. Testing, by other scientists, of the results of an investigation for the purpose of finding or eliminating any errors in reasoning or measurement

   C. Maintaining an open, public process to better promote honesty and integrity in science

   D. Provide a forum to help promote progress through mutual sharing and review of the results of investigations

   Answer: A

   Answers B, C and D all are important rationales for the use of open communication and peer-review in science. Answer A, however, would suggest that the purpose of these processes is to simply maintain the status quo; the history of science, however, suggests that this cannot and should not be the case.

10. **Which of the following devices would be best suited for an experiment designed to measure alpha particle emissions from a sample?**
    (Average Rigor) (Skill 3.1)

    A. Photomultiplier tube

    B. Thermocouple

    C. Geiger-Müller tube

    D. Transistor

    Answer: C

    The Geiger-Müller tube is the main component of the so-called Geiger counter, which is designed specifically for detecting ionizing radiation emissions, including alpha particles. The photomultiplier tube is better suited to measurement of electromagnetic radiation closer to the visible range (A), and the thermocouple is better suited to measurement of temperature (B). Transistors may be involved in instrumentation, but they are not sensors.

11. Which statement best characterizes the relationship of mathematics and experimentation in physics?
(Easy) (Skill 3.2)

    A. Experimentation has no bearing on the mathematical models that are developed for physical phenomena

    B. Mathematics is a tool that assists in the development of models for various physical phenomena as they are studied experimentally, with observations of the phenomena being a test of the validity of the mathematical model

    C. Mathematics is used to test the validity of experimental apparatus for physical measurements

    D. Mathematics is an abstract field with no relationship to concrete experimentation

Answer: B

Mathematics is used extensively in the study of physics for creating models of various phenomena. Since mathematics is abstract and not necessarily tied to physical reality, it must be tempered by experimental results. Although a particular theory may be mathematically elegant, it may have no explanatory power due to its inability to account for certain aspects of physical reality, or due to its inclusion of gratuitous aspects that seem to have no physical analog. Thus, experimentation is foundational, with mathematics being a tool for organizing and providing a greater context for observational results.

12. **What best describes an appropriate view of scale in scientific investigation?**
    (Rigorous) (Skill 3.5)

    A. Scale is irrelevant: the same fundamental principles apply at all scale levels

    B. Scale has no bearing on experimentation, although it may have a part in some theories

    C. Scale has only a practical effect: it may make certain experiments more or less difficult due to limitations on equipment, but it has no fundamental or theoretical impact

    D. Scale is an important consideration both experimentally and theoretically: certain phenomena may have negligible effect at one scale, but may have an overwhelming effect at another scale

Answer: D

Although A is, in a sense, correct, in that the same principles do apply at all scales, their relative effect at various scales can vary drastically. Gravitational and quantum effects each may dominate at different scale levels, for example. As a result, both B and C are incorrect, since scale is a factor both theoretically and experimentally.

13. Which statement best describes a rationale for the use of statistical analysis in characterizing the numerical results of a scientific experiment or investigation?
(Average Rigor) (Skill 3.6)

    A. Experimental results need to be adjusted, through the use of statistics, to conform to theoretical predictions and computer models

    B. Since experiments are prone to a number of errors and uncertainties, statistical analysis provides a method for characterizing experimental measurements by accounting for or quantifying these undesirable effects

    C. Experiments are not able to provide any useful information, and statistical analysis is needed to impose a theoretical framework on the results

    D. Statistical analysis is needed to relate experimental measurements to computer-simulated values

Answer: B

One of the main reasons for the use of statistical analysis is that various types of noise, errors and uncertainties can easily enter into experimental results. Among other things, statistics can help alleviate these difficulties by quantifying an average measurement value and a variance or standard deviation of the set of measurements. This helps determine the accuracy and precision of a set of experimental results. Answers A, C and D do not accurately describe ideal scientific experiments or the use of statistics.

TEACHER CERTIFICATION STUDY GUIDE

14. Which of the following is *not* an SI unit?
    (Average Rigor) (Skill 4.1)

    A. Joule

    B. Coulomb

    C. Newton

    D. Erg

Answer: D

The first three responses are the SI (*Le Système International d'Unités*) units for energy, charge and force, respectively. The fourth answer, the erg, is the CGS (centimeter-gram-second) unit of energy.

15. Which of the following best describes the relationship of precision and accuracy in scientific measurements?
    (Easy) (Skill 4.3)

    A. Accuracy is how well a particular measurement agrees with the value of the actual parameter being measured; precision is how well a particular measurement agrees with the average of other measurements taken for the same value

    B. Precision is how well a particular measurement agrees with the value of the actual parameter being measured; accuracy is how well a particular measurement agrees with the average of other measurements taken for the same value

    C. Accuracy is the same as precision

    D. Accuracy is a measure of numerical error; precision is a measure of human error

Answer: A

The accuracy of a measurement is how close the measurement is to the "true" value of the parameter being measured. Precision is how closely a group of measurements is to the mean value of all the measurements. By analogy, accuracy is how close a measurement is to the center of the bulls-eye, and precision is how tight a group is formed by multiple measurements, regardless of accuracy. Thus, measurements may be very precise and not very accurate, or they may be accurate but not overly precise, or they may be both or neither.

16. Which statement best describes how significant figures must be treated when analyzing a collection of measured data?
    (Rigorous) (Skill 4.4)

    A. Report all data and any results calculated using that data with the number of significant figures that corresponds to the least accurate number (i.e., the one with the fewest significant figures)

    B. Report all data and any results calculated using that data with the number of significant figures that corresponds to the most accurate number (i.e., the one with the most significant figures)

    C. Follow appropriate rules for the propagation of error, maintaining an appropriate number of significant figures for each particular set of data or calculated result

    D. Report all data with the appropriate number of significant figures, but ignore error when calculating results

Answer: C

It is critical, when analyzing data, to maintain the appropriate number of significant figures (a quantification of error) for each individual set of data or calculated result. To do this, certain rules must be followed to appropriately characterize the propagation of error from one or more data values, through various calculations, to the final result. The number of significant figures in the data may be different from the number in a calculated value, such as in an average or variance.

TEACHER CERTIFICATION STUDY GUIDE

17. **Which of the following experiments presents the most likely cause for concern about laboratory safety?**
(Average Rigor) (Skill 4.6)

    A. Computer simulation of a nuclear reactor

    B. Vibration measurement with a laser

    C. Measurement of fluorescent light intensity with a battery-powered photodiode circuit

    D. Ambient indoor ionizing radiation measurement with a Geiger counter.

Answer: B

Assuming no profoundly foolish acts, the use of a computer for simulation (A), measurement with a battery-powered photodiode circuit (C) and ambient radiation measurement (D) pose no particular hazards. The use of a laser (B) must be approached with care, however, as unintentional reflections or a lack of sufficient protection can cause permanent eye damage.

18. **Which of the following mathematical tools would not typically be used for the analysis of an electromagnetic phenomenon?**
(Rigorous) (Skills 5.1-5.5)

    A. Trigonometry

    B. Vector calculus

    C. Group theory

    D. Numerical methods

Answer: C

Trigonometry and vector calculus are both key tools for solving problems in electromagnetics. These are, primarily, analytical methods, although they play a part in numerical analysis as well. Numerical methods are helpful for many problems that are otherwise intractable analytically. Group theory, although it may have some applications in certain highly specific areas, is generally not used in the study of electromagnetics.

TEACHER CERTIFICATION STUDY GUIDE

19. For a problem that involves parameters that vary in rate with direction and location, which of the following mathematical tools would most likely be of greatest value?
    (Rigorous) (Skill 5.5)

    A. Trigonometry

    B. Numerical analysis

    C. Group theory

    D. Vector calculus

    Answer: D

    Each of the above answers might have some value for individual problems, but, generally speaking, those problems that deal with quantities that have direction and magnitude (vectors), and that deal with rates, would most likely be amenable to analysis using vector calculus (D).

20. When acceleration is plotted versus time, the area under the graph represents:
    (Average Rigor) (Skill 6.3)

    A. Time

    B. Distance

    C. Velocity

    D. Acceleration

    Answer: C

    The area under a graph will have units equal to the product of the units of the two axes. (To visualize this, picture a graphed rectangle with its area equal to length times width.)
    Therefore, multiply units of acceleration by units of time:
    (length/time^2)(time)
    This equals length/time, i.e. units of velocity.

21. A brick and hammer fall from a ledge at the same time. They would be expected to:
(Easy) (Skill 6.4)

    A. Reach the ground at the same time

    B. Accelerate at different rates due to difference in weight

    C. Accelerate at different rates due to difference in potential energy

    D. Accelerate at different rates due to difference in kinetic energy

Answer: A

This is a classic question about falling in a gravitational field. All objects are acted upon equally by gravity, so they should reach the ground at the same time. (In real life, air resistance can make a difference, but not at small heights for similarly shaped objects.) In any case, weight, potential energy, and kinetic energy do not affect gravitational acceleration. Thus, the only possible answer is (A).

22. Given a vase full of water, with holes punched at various heights. The water squirts out of the holes, achieving different distances before hitting the ground. Which of the following accurately describes the situation?
(Average Rigor) (Skill 7.6)

    A. Water from higher holes goes farther, due to Pascal's Principle.

    B. Water from higher holes goes farther, due to Bernoulli's Principle.

    C. Water from lower holes goes farther, due to Pascal's Principle.

    D. Water from lower holes goes farther, due to Bernoulli's Principle.

Answer: D

To answer this question, consider the pressure on the water in the vase. The deeper the water, the higher the pressure. Thus, when a hole is punched, the water stream will achieve higher velocity as it equalizes to atmospheric pressure. The lower streams will therefore travel farther before hitting the ground. This eliminates answers (A) and (B). Then recall that Pascal's Principle provides for immediate pressure changes throughout a fluid, while Bernoulli's Principle translates pressure, velocity, and height energy into each other. In this case, the pressure energy is being transformed into velocity energy, and Bernoulli's Principle applies. Therefore, the only appropriate answer is (D).

TEACHER CERTIFICATION STUDY GUIDE

23. The magnitude of a force is:
    (Easy) (Skill 8.1)

    A. Directly proportional to mass and inversely to acceleration

    B. Inversely proportional to mass and directly to acceleration

    C. Directly proportional to both mass and acceleration

    D. Inversely proportional to both mass and acceleration

Answer: C

To solve this problem, recall Newton's 2nd Law, i.e. net force is equal to mass times acceleration. Therefore, the only possible answer is (C).

24. If a force of magnitude *F* gives a mass *M* an acceleration *A*, then a force 3*F* would give a mass 3*M* an acceleration:
    (Average Rigor) (Skill 8.2)

    A. *A*

    B. 12*A*

    C. *A*/2

    D. 6*A*

Answer: A

To solve this problem, apply Newton's Second Law, which is also implied by the first part of the problem:
Force = (Mass)(Acceleration)
F = MA
Then apply the same law to the second case, and isolate the unknown:
3*F* = 3*M* x
x = (3*F*)/(3*M*)
x = *F*/*M*
x = *A* (by substituting from our first equation)
Only answer (A) matches these calculations.

PHYSICS

25. An inclined plane is tilted by gradually increasing the angle of elevation θ, until the block will slide down at a constant velocity. The coefficient of friction, μ_k, is given by:
    (Rigorous) (Skill 8.3)

    A. cos θ

    B. sin θ

    C. cosecant θ

    D. tangent θ

Answer: D

When the block moves, its force upstream (due to friction) must equal its force downstream (due to gravity).

The friction force is given by
$F_f = \mu_k N$
where μ_k is the friction coefficient and N is the normal force.

Using similar triangles, the gravity force is given by
$F_g = mg \sin \theta$
and the normal force is given by
$N = mg \cos \theta$

When the block moves at constant velocity, it must have zero net force, so set equal the force of gravity and the force due to friction:
$F_f = F_g$
$\mu_k mg \cos \theta = mg \sin \theta$
$\mu_k = \tan \theta$

Answer (D) is the only appropriate choice in this case.

TEACHER CERTIFICATION STUDY GUIDE

26. A mass of 2 kg connected to a spring undergoes simple harmonic motion at a frequency of 3 Hz. What is the spring constant?
    (Average Rigor) (Skill 8.4)

    A. 6 kg/s^2

    B. 18 kg/s^2

    C. 710 kg/s^2

    D. 1000 kg/s^2

Answer: C

The spring constant, k, is equal to mω^2. In this case, ω is equal to 2π times the frequency of 3 Hz. The spring constant may be derived quickly by recognizing that the position of the mass varies sinusoidally with time at an angular frequency ω. Noting that the acceleration is the second derivative of the position with respect to time, the expression for k in Hooke's law (F = -kx) can be easily derived.

27. A projectile with a mass of 1.0 kg has a muzzle velocity of 1500.0 m/s when it is fired from a cannon with a mass of 500.0 kg. If the cannon slides on a frictionless track, it will recoil with a velocity of ____ m/s.
    (Rigorous) (Skill 8.5)

    A. 2.4

    B. 3.0

    C. 3.5

    D. 1500

Answer: B

To solve this problem, apply Conservation of Momentum to the cannon-projectile system. The system is initially at rest, with total momentum of 0 kg m/s. Since the cannon slides on a frictionless track, we can assume that the net momentum stays the same for the system. Therefore, the momentum forward (of the projectile) must equal the momentum backward (of the cannon). Thus:

$p_{projectile} = p_{cannon}$
$m_{projectile} \, v_{projectile} = m_{cannon} \, v_{cannon}$
(1.0 kg)(1500.0 m/s) = (500.0 kg)(x)
x = 3.0 m/s

Only answer (B) matches these calculations.

28. If an object of length 5 meters (along the ŷ axis) is traveling at 99% of the speed of light in the x̂ direction, what is its length, as measured by an observer at the origin?
(Rigorous) (Skill 8.6)

    A. 0 m

    B. 0.71 m

    C. 3.4 m

    D. 5 m

Answer: D

This problem, since it involves a speed close to c, must be treated relativistically. The answer may be quickly determined, however, by realizing that there is no observed length contraction perpendicular to the direction of travel. Thus, if an object is moving in the x direction, only its x dimension is contracted. As a result, there is no need to calculate the Lorentz factor; the answer is simply the proper length of the object (5 meters).

29. An object traveling through air loses part of its energy of motion due to friction. Which statement best describes what has happened to this energy?
(Easy) (Skill 9.1)

    A. The energy is destroyed

    B. The energy is converted to static charge

    C. The energy is radiated as electromagnetic waves

    D. The energy is lost to heating of the air

Answer: D

Since energy must be conserved, the energy of motion of the object is converted, in part, to energy of motion of the molecules in the air (and, to some extent, in the object). This additional motion is equivalent to an increase in heat. Thus, friction is a loss of energy of motion through heating.

TEACHER CERTIFICATION STUDY GUIDE

30. A 100 g mass revolving around a fixed point, on the end of a 0.5 meter string, circles once every 0.25 seconds. What is the magnitude of the centripetal acceleration?
(Average Rigor) (Skill 10.2)

   A. 1.23 m/s^2

   B. 31.6 m/s^2

   C. 100 m/s^2

   D. 316 m/s^2

Answer: D

The centripetal acceleration is equal to the product of the radius and the square of the angular frequency ω. In this case, ω is equal to 25.1 Hz. Squaring this value and multiplying by 0.5 m yields the result in answer D.

31. A satellite is in a circular orbit above the earth. Which statement is false?
(Average Rigor) (Skill 10.3)

   A. An external force causes the satellite to maintain orbit.

   B. The satellite's inertia causes it to maintain orbit.

   C. The satellite is accelerating toward the earth.

   D. The satellite's velocity and acceleration are not in the same direction.

Answer: B

To answer this question, recall that in circular motion, an object's inertia tends to keep it moving straight (tangent to the orbit), so a centripetal force (leading to centripetal acceleration) must be applied. In this case, the centripetal force is gravity due to the earth, which keeps the object in motion. Thus, (A), (C), and (D) are true, and (B) is the only false statement.

TEACHER CERTIFICATION STUDY GUIDE

32. A simple pendulum with a period of one second has its mass doubled. If the length of the string is quadrupled, the new period will be: *(Rigorous) (Skill 10.4)*

    A. 1 second

    B. 2 seconds

    C. 3 seconds

    D. 5 seconds

Answer: B

To answer this question, recall that the period of a pendulum is given by:

$T = 2\pi (L/g)^{1/2}$ where T is period; L is length; g is gravitational acceleration (This is derived from balancing the forces and making small-angle approximation for small angles.)

Note that this equation is independent of mass, so that change is irrelevant. Since the length is quadrupled, and all other quantities on the right side of the equation are constant, the new period will be increased by a factor of two (the square root of four).

This is consistent only with answer (B).

33. A uniform pole weighing 100 grams, that is one meter in length, is supported by a pivot at 40 centimeters from the left end. In order to maintain static position, a 200 gram mass must be placed _____ centimeters from the left end.
(Rigorous) (Skill 10.6)

    A. 10

    B. 45

    C. 35

    D. 50

Answer: C

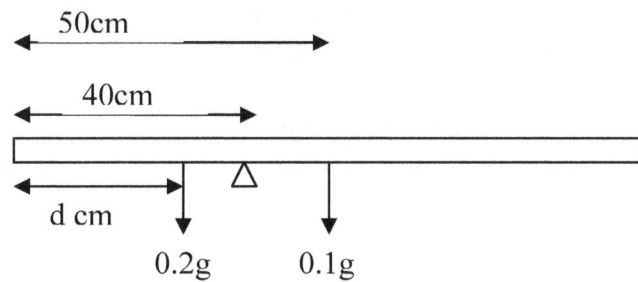

Since the pole is uniform, we can assume that its weight 0.1g acts at the center, i.e. 50 cm from the left end. In order to keep the pole balanced on the pivot, the 200 gram mass must be placed such that the torque on the pole due to the mass is equal and opposite to the torque due to the pole's weight. Thus, if the 200 gram mass is placed d cm from the left end of the pole,

$(40 - d) \times 0.2g = 10 \times 0.1g$; $40 - d = 5$; $d = 35$ cm

TEACHER CERTIFICATION STUDY GUIDE

34. Which of the following units is not used to measure torque?
    (Average Rigor) (Skill 10.7)

    A. slug ft

    B. lb ft

    C. N m

    D. dyne cm

Answer: A

To answer this question, recall that torque is always calculated by multiplying units of force by units of distance. Therefore, answer (A), which is the product of units of mass and units of distance, must be the choice of incorrect units. Indeed, the other three answers all could measure torque, since they are of the correct form. It is a good idea to review "English Units" before the teacher test, because they are occasionally used in problems.

35. A gas maintained at a constant pressure has a specific heat which is greater than its specific heat when maintained at constant volume, because:
(Rigorous) (Skill 11.1)

    A. The Coefficient of Expansion changes.

    B. The gas enlarges as a whole.

    C. Brownian motion causes random activity.

    D. Work is done to expand the gas.

Answer: D

To answer this question, recall that the specific heat is a measure of how much energy it takes to raise the temperature of a given mass of gas. Thus, you can reason that when a gas is maintained at constant pressure, some energy is used to expand the volume of the gas, and less is left for temperature changes. In fact, this is the case, and (D) is the correct answer. If you were not able to figure that out, you could still eliminate the other answers, because they are not strictly relevant to a change in specific heat.

36. The First Law of Thermodynamics takes the form dU = dW when the conditions are:
 (Rigorous) (Skill 11.2)

   A. Isobaric

   B. Isochloremic

   C. Isothermal

   D. Adiabatic

Answer: D

To answer this question, recall the First Law of Thermodynamics:
Change in Internal Energy = Work Done + Heat Added
dU = dW + dQ

Thus in the form we are given, dQ has been set to zero, i.e. there is no heat added. "Adiabatic" refers to a case where there is no heat exchange with surroundings, so answer (D) is the appropriate choice. "Isobaric" means at a constant pressure, "Isothermal" means at a constant temperature, and "Isochloremic" is an imaginary word, as far as I can tell.

It might be tempting to choose "Isothermal," thinking that no heat added would require the same temperature. However, work and internal energy changes can change temperature within the system analyzed, even when no heat is exchanged with the surroundings.

TEACHER CERTIFICATION STUDY GUIDE

37. A temperature change of 40 degrees Celsius is equal to a change in Fahrenheit degrees of:
(Average Rigor) (Skill 11.4)

    A. 40

    B. 20

    C. 72

    D. 112

Answer: C

To answer this question, recall the equation for Celsius and Fahrenheit:
°F = 9/5 °C + 32

Therefore, whatever temperature difference occurs in °C, it is multiplied by a factor of 9/5 to get the new °F measurement:
new°F = 9/5(old°C + 40) + 32
(whereas old°F = 9/5(old°C) + 32)
Therefore the difference between the old and new temperatures in Fahrenheit is 9/5 of 40, or 72 degrees. This is consistent only with answer (C).

38. A cooking thermometer in an oven works because the metals it is composed of have different:
(Average Rigor) (Skill 11.5)

    A. Melting points

    B. Heat convection

    C. Magnetic fields

    D. Coefficients of expansion

Answer: D

A thermometer of the type that can withstand oven temperatures works by having more than one metal strip. These strips expand at different rates with temperature increases, causing the dial to register the new temperature. This is consistent only with answer (D). If you did not know how an oven thermometer works, you could still omit the incorrect answers: It is unlikely that the metals in a thermometer would melt in the oven to display the temperature; the magnetic fields would not be useful information in this context; heat convection applies in fluids, not solids.

TEACHER CERTIFICATION STUDY GUIDE

39. **Heat transfer by electromagnetic waves is termed:**
(Easy) (Skill 11.6)

    A. Conduction

    B. Convection

    C. Radiation

    D. Phase Change

Answer: C

To answer this question, recall the different ways that heat is transferred. Conduction is the transfer of heat through direct physical contact and molecules moving and hitting each other. Convection is the transfer of heat via density differences and flow of fluids. Radiation is the transfer of heat via electromagnetic waves (and can occur in a vacuum). Phase Change causes transfer of heat (though not of temperature) in order for the molecules to take their new phase. This is consistent, therefore, only with answer (C).

40. **A long copper bar has a temperature of 60°C at one end and 0°C at the other. The bar reaches thermal equilibrium (barring outside influences) by the process of heat:**
(Average Rigor) (Skill 11.6)

    A. Fusion

    B. Convection

    C. Conduction

    D. Microwaving

Answer: C

To answer this question, recall the different methods of heat transfer. (Note that since the bar is warm at one end and cold at the other, heat must transfer through the bar from warm to cold, until temperature is equalized.) "Convection" is the heat transfer via fluid currents. "Conduction" is the heat transfer via connected solid material. "Fusion" and "Microwaving" are not methods of heat transfer. Therefore the only appropriate answer is (C).

41. Use the information on heats below to solve this problem. An ice block at 0° Celsius is dropped into 100 g of liquid water at 18° Celsius. When thermal equilibrium is achieved, only liquid water at 0° Celsius is left. What was the mass, in grams, of the original block of ice?

Given: Heat of fusion of ice = 80 cal/g
Heat of vaporization of ice = 540 cal/g
Specific Heat of ice = 0.50 cal/g°C
Specific Heat of water = 1 cal/g°C

(Rigorous) (Skill 11.7)

A. 2.0

B. 5.0

C. 10.0

D. 22.5

Answer: D

To solve this problem, apply Conservation of Energy to the ice-water system. Any gain of heat to the melting ice must be balanced by loss of heat in the liquid water. Use the two equations relating temperature, mass, and energy:
$Q = m C \Delta T$ (for heat loss/gain from change in temperature)
$Q = m L$ (for heat loss/gain from phase change)
where Q is heat change; m is mass; C is specific heat; ΔT is change in temperature; L is heat of phase change (in this case, melting, also known as "fusion").

Then
$Q_{\text{ice to water}} = Q_{\text{water to ice}}$
(Note that the ice only melts; it stays at 0° Celsius—otherwise, we would have to include a term for warming the ice as well. Also the information on the heat of vaporization for water is irrelevant to this problem.)
$m L = m C \Delta T$
$x (80 \text{ cal/g}) = 100g \cdot 1 \text{cal/g°C} \cdot 18°C$
$x (80 \text{ cal/g}) = 1800 \text{ cal}$
$x = 22.5 \text{ g}$

Only answer (D) matches this result.

42. **Which of the following is not an assumption upon which the kinetic-molecular theory of gases is based?**
    (Rigorous) (Skill 11.8)

    A. Quantum mechanical effects may be neglected

    B. The particles of a gas may be treated statistically

    C. The particles of the gas are treated as very small masses

    D. Collisions between gas particles and container walls are inelastic

Answer: D

Since the kinetic-molecular theory is classical in nature, quantum mechanical effects are indeed ignored, and answer A is incorrect. The theory also treats gases as a statistical collection of point-like particles with finite masses. As a result, answers B and C may also be eliminated. Thus, answer D is correct: collisions between gas particles and container walls are treated as elastic in the kinetic-molecular theory.

43. What should be the behavior of an electroscope, which has been grounded in the presence of a positively charged object (1), after the ground connection is removed and then the charged object is removed from the vicinity (2)?
(Average Rigor) (Skill 12.1)

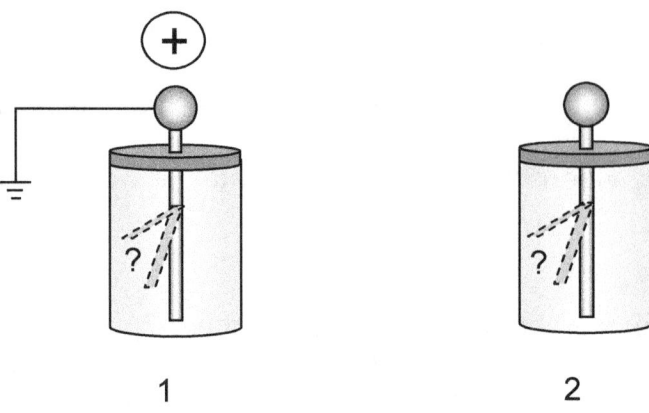

A. The metal leaf will start deflected (1) and then relax to an undeflected position (2)

B. The metal leaf will start in an undeflected position (1) and then be deflected (2)

C. The metal leaf will remain undeflected in both cases

D. The metal leaf will be deflected in both cases

Answer: B

When grounded, the electroscope will show no deflection. Nevertheless, if the ground is then removed and the charged object taken from the vicinity (in that order), the excess charge that existed near the sphere of the electroscope will distribute itself throughout the instrument, resulting in an overall net excess charge that will deflect the metal leaf.

44. Static electricity generation occurs by:
    (Easy) (Skill 12.2)

    A. Telepathy

    B. Friction

    C. Removal of heat

    D. Evaporation

Answer: B

Static electricity occurs because of friction and electric charge build-up. There is no such thing as telepathy, and neither removal of heat nor evaporation are causes of static electricity. Therefore, the only possible answer is (B).

45. The electric force in Newtons, on two small objects (each charged to – 10 microCoulombs and separated by 2 meters) is:
    (Rigorous) (Skill 12.6)

    A. 1.0

    B. 9.81

    C. 31.0

    D. 0.225

Answer: D

To answer this question, use Coulomb's Law, which gives the electric force between two charged particles:
$F = k Q_1 Q_2 / r^2$
Then our unknown is F, and our knowns are:
$k = 9.0 \times 10^9$ Nm^2/C^2
$Q_1 = Q_2 = -10 \times 10^{-6}$ C
$r = 2$ m

Therefore
$F = (9.0 \times 10^9)(-10 \times 10^{-6})(-10 \times 10^{-6})/(2^2)$ N
$F = 0.225$ N

This is compatible only with answer (D).

PHYSICS

TEACHER CERTIFICATION STUDY GUIDE

46. When the current flowing through a fixed resistance is doubled, the amount of heat generated is:
    (Average Rigor) (Skill 13.1 and 13.4)

    A. Quadrupled

    B. Doubled

    C. Multiplied by pi

    D. Halved

Answer: A

To answer this question, recall that heat generated will occur because of the power of the circuit (power is energy per time). For a circuit with a fixed resistance:
$$P = I V$$
where P is power; I is current; V is voltage. Then use Ohm's Law:
$$V = I R$$
where V is voltage; I is current; R is resistance, and substitute:
$$P = I^2 R$$
and so the doubling of the current I will lead to a quadrupling of the power, and therefore the a quadrupling of the heat.

This is consistent only with answer (A). If you weren't sure of the equations, you could still deduce that with more current, there would be more heat generated, and therefore eliminate answer choice (D) in any case.

47. A semi-conductor allows current to flow:
    (Easy) (Skill 13.3)

    A. Never

    B. Always

    C. As long as it stays below a maximum temperature

    D. When a minimum voltage is applied

Answer: D

To answer this question, recall that semiconductors do not conduct as well as conductors (eliminating answer (B)), but they conduct better than insulators (eliminating answer (A)). Semiconductors can conduct better when the temperature is higher (eliminating answer (C)), and their electrons move most readily under a potential difference. Thus the answer can only be (D).

PHYSICS

48. A 10 ohm resistor and a 50 ohm resistor are connected in parallel. If the current in the 10 ohm resistor is 5 amperes, the current (in amperes) running through the 50 ohm resistor is:
(Rigorous) (Skills 13.1 and 13.4)

    A. 1

    B. 50

    C. 25

    D. 60

Answer: A

To answer this question, use Ohm's Law, which relates voltage to current and resistance:
V = IR
where V is voltage; I is current; R is resistance.

We also use the fact that in a parallel circuit, the voltage is the same across the branches.

Because we are given that in one branch, the current is 5 amperes and the resistance is 10 ohms, we deduce that the voltage in this circuit is their product, 50 volts (from V = IR).

We then use V = IR again, this time to find I in the second branch. Because V is 50 volts, and R is 50 ohm, we calculate that I has to be 1 ampere.

This is consistent only with answer (A).

49. How much power is dissipated through the following resistive circuit? *(Average Rigor) (Skill 13.6 and 13.7)*

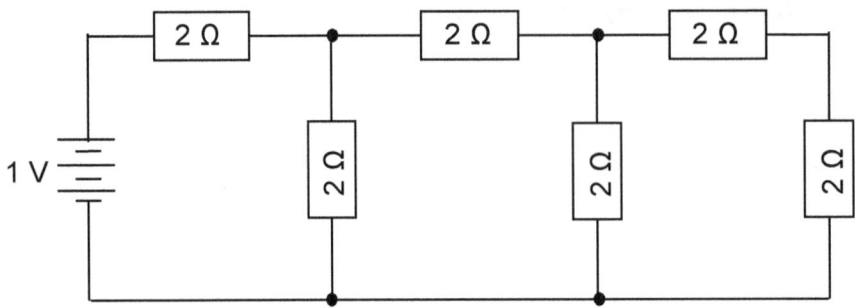

A. 0 W

B. 0.22 W

C. 0.31 W

D. 0.49 W

Answer: C

Use the rules of series and parallel resistors to quickly form an equivalent circuit with a single voltage source and a single resistor. In this case, the equivalent resistance is 3.25 Ω. The power dissipated by the circuit is the square of the voltage divided by the resistance. The final answer is C.

50. The greatest number of 100 watt lamps that can be connected in parallel with a 120 volt system without blowing a 5 amp fuse is:
(Rigorous) (Skill 13.7)

    A. 24

    B. 12

    C. 6

    D. 1

Answer: C

To solve fuse problems, you must add together all the drawn current in the parallel branches, and make sure that it is less than the fuse's amp measure. Because we know that electrical power is equal to the product of current and voltage, we can deduce that:
$I = P/V$ (I = current (amperes); P = power (watts); V = voltage (volts))

Therefore, for each lamp, the current is 100/120 amperes, or 5/6 ampere. The highest possible number of lamps is thus six, because six lamps at 5/6 ampere each adds to 5 amperes; more will blow the fuse.

This is consistent only with answer (C).

51. What is the electric flux density (or electric displacement) through each face of a cube, of side length 2 meters, that contains a central point charge of 2 Coulombs?
(Rigorous) (Skill 13.8)

    A. 0.50 Coulomb/m^2

    B. 0.33 Coulomb/m^2

    C. 0.13 Coulomb/m^2

    D. 0.33 Tesla

Answer: B

Since the point charge is located at the center of the cube, a symmetric situation exists. Furthermore, since electric flux density has the same value regardless of the shape or size of the surface that encloses the charge, the size of the cube is irrelevant. The amount of flux, therefore, is the same through each face of the cube. Simply divide the charge by the number of cube faces (six) to obtain the result in answer B.

52. All of the following use semi-conductor technology, except a(n):
(Average Rigor) (Skill 13.9)

   A. Transistor

   B. Diode

   C. Capacitor

   D. Operational Amplifier

Answer: C

Semi-conductor technology is used in transistors and operational amplifiers, and diodes are the basic unit of semi-conductors. Therefore the only possible answer is (C), and indeed a capacitor does not require semi-conductor technology.

53. What is the direction of the magnetic field at the center of the loop of current (I) shown below (i.e., at point A)?
(Easy) (Skill 14.1)

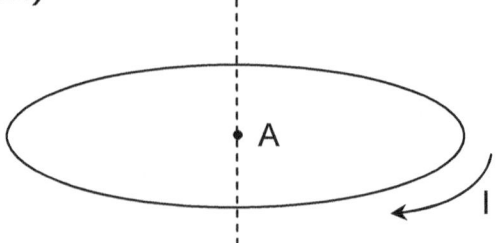

   A. Down, along the axis (dotted line)

   B. Up, along the axis (dotted line)

   C. The magnetic field is oriented in a radial direction

   D. There is no magnetic field at point A

Answer: A

The magnetic field may be found by applying the right-hand rule. The magnetic field curls around the wire in the direction of the curled fingers when the thumb is pointed in the direction of the current. Since there is a degree of symmetry, with point A lying in the center of the loop, the contributions of all the current elements on the loop must yield a field that is either directed up or down at the axis. Use of the right-hand rule indicates that the field is directed down. Thus, answer A is correct.

54. **What effect might an applied external magnetic field have on the magnetic domains of a ferromagnetic material?**
(Rigorous) (Skill 14.2)

> A. The domains that are not aligned with the external field increase in size, but those that are aligned decrease in size
>
> B. The domains that are not aligned with the external field decrease in size, but those that are aligned increase in size
>
> C. The domains align perpendicular to the external field
>
> D. There is no effect on the magnetic domains

Answer: B

Recall that ferromagnetic domains are portions of a magnetic material that have a local magnetic moment. The material may have an overall lack of a magnetic moment due to random alignment of its domains. In the presence of an applied field, the domains may align with the field to some extent, or the boundaries of the domains may shift to give greater weight to those domains that are aligned with the field, at the expense of those domains that are not aligned with the field. As a result, of the possibilities above, B is the best answer.

55. **Which of the following statements may be taken as a legitimate inference based upon the Maxwell equation that states $\nabla \cdot \mathbf{B} = 0$?**
(Average Rigor) (Skill 14.6)

> A. The electric and magnetic fields are decoupled
>
> B. The electric and magnetic fields are mediated by the W boson
>
> C. There are no photons
>
> D. There are no magnetic monopoles

Answer: D

Since the divergence of the magnetic flux density is always zero, there cannot be any magnetic monopoles (charges), given this Maxwell equation. If Gauss's law is applied to magnetic flux in the same manner as it is to electric flux, then the total magnetic "charge" contained within any closed surface must always be zero. This is another way of viewing the problem. Thus, answer D is correct. This answer may also be chosen by elimination of the other statements, which are untenable.

TEACHER CERTIFICATION STUDY GUIDE

56. The use of two circuits next to each other, with a change in current in the primary circuit, demonstrates:
    (Rigorous) (Skill 15.1)

    A. Mutual current induction

    B. Dielectric constancy

    C. Harmonic resonance

    D. Resistance variation

Answer: A

To answer this question, recall that changing current induces a change in magnetic flux, which in turn causes a change in current to oppose that change (Lenz's and Faraday's Laws). Thus, (A) is correct. If you did not remember that, note that harmonic resonance is irrelevant here (eliminating (C)), and there is no change in resistance in the circuits (eliminating (D)).

57. A static magnetic flux density of 1 Tesla is linked by a wire loop, as shown below. What is the electromotive force (EMF) around the loop? *(Average Rigor) (Skill 15.1)*

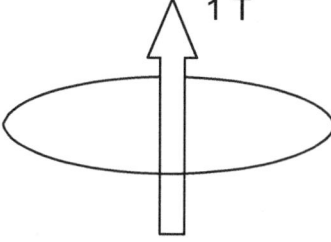

A. 1 Volt

B. 2 Volts

C. 5 Volts

D. 0 Volts

Answer: D

According to Maxwell's equations, the curl of the electric field is proportional to the time rate of change of the magnetic field.

$$\nabla \times \mathbf{E} = -\frac{\partial \mathbf{B}}{\partial t}$$

Since the situation above is source-free, the divergence of the electric field is zero. Since, as noted in the question, the magnetic flux is static, the variation of the magnetic field with time must also be zero, as must the curl of the electric field. Since the curl and divergence of the electric field are both zero, the field itself must be zero, and there can be no potential difference around the wire. Put simply, only when there is a change in the magnetic flux through the wire loop is there a corresponding EMF.

58. What is the effect of running current in the same direction along two parallel wires, as shown below?
(Rigorous) (Skill 15.2)

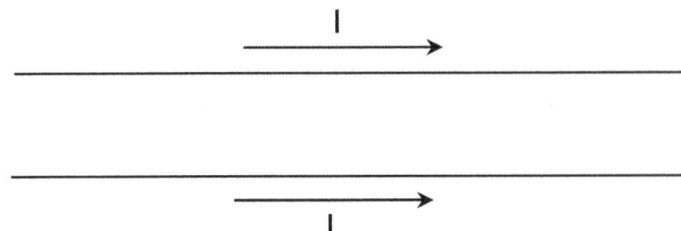

A. There is no effect

B. The wires attract one another

C. The wires repel one another

D. A torque is applied to both wires

Answer: B

Since the direction of the force on a current element is proportional to the cross product of the direction of the current element and the magnetic field, there is either an attractive or repulsive force between the two wires shown above. Using the right hand rule, it can be found that the magnetic field on the top wire due to the bottom wire is directed out of the plane of the page. Performing the cross product shows that the force on the upper wire is directed toward the lower wire. A similar argument can be used for the lower wire. Thus, the correct answer is B: an attractive force is exerted on the wires.

TEACHER CERTIFICATION STUDY GUIDE

59. A wave generator is used to create a succession of waves. The rate of wave generation is one every 0.33 seconds. The period of these waves is:
(Average Rigor) (Skills 16.1 and 16.3)

   A. 2.0 seconds

   B. 1.0 seconds

   C. 0.33 seconds

   D. 3.0 seconds

Answer: C

The definition of a period is the length of time between wave crests. Therefore, when waves are generated one per 0.33 seconds, that same time (0.33 seconds) is the period. This is consistent only with answer (C). Do not be trapped into calculating the number of waves per second, which might lead you to choose answer (D).

60. An electromagnetic wave propagates through a vacuum. Independent of its wavelength, it will move with constant:
(Easy) (Skill 16.2)

   A. Acceleration

   B. Velocity

   C. Induction

   D. Sound

Answer: B

Electromagnetic waves are considered always to travel at the speed of light, so answer (B) is correct. Answers (C) and (D) can be eliminated in any case, because induction is not relevant here, and sound does not travel in a vacuum.

TEACHER CERTIFICATION STUDY GUIDE

61. A wave has speed 60 m/s and wavelength 30,000 m. What is the frequency of the wave?
(Average Rigor) (Skill 16.3)

    A. 2.0×10^{-3} Hz

    B. 60 Hz

    C. 5.0×10^{2} Hz

    D. 1.8×10^{6} Hz

Answer: A

To answer this question, recall that wave speed is equal to the product of wavelength and frequency. Thus:
$$60 \text{ m/s} = (30,000 \text{ m}) \text{ (frequency)}$$
$$\text{frequency} = 2.0 \times 10^{-3} \text{ Hz}$$

This is consistent only with answer (A).

62. The wave phenomenon of polarization applies only to:
(Average Rigor) (Skill 16.5)

    A. Longitudinal waves

    B. Transverse waves

    C. Sound

    D. Light

Answer: B

To answer this question, recall that polarization is when waves are screened so that they come out aligned in a certain direction. (To illustrate this, take two pairs of polarizing sunglasses, and note the light differences when rotating one lens over another. When the lenses are polarizing perpendicularly, no light gets through.) This applies only to transverse waves, which have wave parts to align. Light can be polarized, but it is not the only wave that can be. Thus, the correct answer is (B).

TEACHER CERTIFICATION STUDY GUIDE

63. If one sound is ten decibels louder than another, the ratio of the intensity of the first to the second is:
(Average Rigor) (Skill 17.1)

    A. 20:1

    B. 10:1

    C. 1:1

    D. 1:10

Answer: B

To answer this question, recall that a decibel is defined as ten times the log of the ratio of sound intensities:
(decibel measure) = 10 log (I / I_0) where I_0 is a reference intensity.

Therefore, in our case,
(decibels of first sound) = (decibels of second sound) + 10
10 log (I_1 / I_0) = 10 log (I_2 / I_0) + 10
10 log I_1 − 10 log I_0 = 10 log I_2 − 10 log I_0 + 10
10 log I_1 − 10 log I_2 = 10
log (I_1 / I_2) = 1
I_1 / I_2 = 10

This is consistent only with answer (B).
(Be careful not to get the two intensities confused with each other.)

TEACHER CERTIFICATION STUDY GUIDE

64. **The velocity of sound is greatest in:**
    (Average Rigor) (Skill 17.2)

    A. Water

    B. Steel

    C. Alcohol

    D. Air

Answer: B

Sound is a longitudinal wave, which means that it shakes its medium in a way that propagates as sound traveling. The speed of sound depends on both elastic modulus and density, but for a comparison of the above choices, the answer is always that sound travels faster through a solid like steel, than through liquids or gases. Thus, the answer is (B).

65. **Which of the following apparatus can be used to measure the wavelength of a sound produced by a tuning fork?**
    (Average Rigor) (Skill 17.3)

    A. A glass cylinder, some water, and iron filings

    B. A glass cylinder, a meter stick, and some water

    C. A metronome and some ice water

    D. A comb and some tissue

Answer: B

To answer this question, recall that a sound will be amplified if it is reflected back to cause positive interference. This is the principle behind musical instruments that use vibrating columns of air to amplify sound (e.g. a pipe organ).

Therefore, presumably a person could put varying amounts of water in the cylinder, and hold the vibrating tuning fork above the cylinder in each case. If the tuning fork sound is amplified when put at the top of the column, then the length of the air space would be an integral multiple of the sound's wavelength. This experiment is consistent with answer (B). Although the experiment would be tedious, none of the other options for materials suggest a better alternative.

PHYSICS

66. A vibrating string's frequency is _____ proportional to the _____.
(Rigorous) (Skill 17.4)

     A. Directly; Square root of the tension

     B. Inversely; Length of the string

     C. Inversely; Squared length of the string

     D. Inversely; Force of the plectrum

Answer: A

To answer this question, recall that
$f = (n v) / (2 L)$ where f is frequency; v is velocity; L is length

and

$v = (F_{tension} / (m / L))^{½}$ where $F_{tension}$ is tension; m is mass; others as above

so

$f = (n / 2 L) ((F_{tension} / (m / L))^{½})$

indicating that frequency is directly proportional to the square root of the tension force. This is consistent only with answer (A). Note that in the final frequency equation, there is an inverse relationship with the square root of the length (after canceling like terms). This is not one of the options, however.

TEACHER CERTIFICATION STUDY GUIDE

67. The highest energy is associated with:
(Easy) (Skill 18.1)

    A. UV radiation

    B. Yellow light

    C. Infrared radiation

    D. Gamma radiation

Answer: D

To answer this question, recall the electromagnetic spectrum. The highest energy (and therefore frequency) rays are those with the lowest wavelength, i.e. gamma rays. (In order of frequency from lowest to highest are: radio, microwave, infrared, red through violet visible light, ultraviolet, X-rays, gamma rays.) Thus, the only possible answer is (D). Note that even if you did not remember the spectrum, you could deduce that gamma radiation is considered dangerous and thus might have the highest energy.

68. In order to switch between two different reference frames in special relativity, we use the _____ transformation.
(Average Rigor) (Skill 18.5)

    A. Galilean

    B. Lorentz

    C. Euclidean

    D. Laplace

Answer: B

The Lorentz transformation is the set of equations to scale length and time between inertial reference frames in special relativity, when velocities are close to the speed of light.

The Galilean transformation is a parallel set of equations, used for 'classical' situations when velocities are much slower than the speed of light. Euclidean geometry is useful in physics, but not relevant here. Laplace transforms are a method of solving differential equations by using exponential functions. The correct answer is therefore (B).

69. A crew is on-board a spaceship, traveling at 60% of the speed of light with respect to the earth. The crew measures the length of their ship to be 240 meters. When a ground-based crew measures the apparent length of the ship, it equals:
(Rigorous) (Skill 18.5)

      A. 400 m

      B. 300 m

      C. 240 m

      D. 192 m

Answer: D

To answer this question, recall that a moving object's size seems contracted to the stationary observer, according to the equation:

$$\text{Length Observed} = (\text{Actual Length})(1 - v^2/c^2)^{1/2}$$

where v is speed of motion, and c is speed of light.

Therefore, in this case,
$$\text{Length Observed} = (240 \text{ m})(1 - 0.6^2)^{1/2}$$
$$\text{Length Observed} = (240 \text{ m})(0.8) = 192 \text{ m}$$

This is consistent only with answer (D). If you were unsure of the equation, you could still reason that because of length contraction (the flip side of time dilation), you must choose an answer with a smaller length, and only (D) fits that description. Note that only the dimension in the direction of travel is contracted. (The length in this case.)

70. All of the following phenomena are considered "refractive effects" except for:
    (Rigorous) (Skills 16.5 and 18.6)

    A. The red shift

    B. Total internal reflection

    C. Lens dependent image formation

    D. Snell's Law

Answer: A

Refractive effects are phenomena that are related to or caused by refraction. The red shift refers to the Doppler Effect as applied to light when galaxies travel away from observers. Total internal reflection is when light is totally reflected in a substance, with no refracted ray into the substance beyond (e.g. in fiber optic cables). It occurs because of the relative indices of refraction in the materials. Lens dependent image formation refers to making images depending on the properties (including index of refraction) of the lens. Snell's Law provides a mathematical relationship for angles of incidence and refraction. Therefore, the only possible answer is (A).

TEACHER CERTIFICATION STUDY GUIDE

71. A stationary sound source produces a wave of frequency *F*. An observer at position A is moving toward the horn, while an observer at position B is moving away from the horn. Which of the following is true?
(Rigorous) (Skill 18.6)

    A. $F_A < F < F_B$

    B. $F_B < F < F_A$

    C. $F < F_A < F_B$

    D. $F_B < F_A < F$

Answer: B

To answer this question, recall the Doppler Effect. As a moving observer approaches a sound source, s/he intercepts wave fronts sooner than if s/he were standing still. Therefore, the wave fronts seem to be coming more frequently. Similarly, as an observer moves away from a sound source, the wave fronts take longer to reach him/her. Therefore, the wave fronts seem to be coming less frequently. Because of this effect, the frequency at B will seem lower than the original frequency, and the frequency at A will seem higher than the original frequency. The only answer consistent with this is (B). Note also, that even if you weren't sure of which frequency should be greater/smaller, you could still reason that A and B should have opposite effects, and be able to eliminate answer choices (C) and (D).

TEACHER CERTIFICATION STUDY GUIDE

72. Automobile mirrors that have a sign, "objects are closer than they appear" say so because:
(Rigorous) (Skill 19.1)

  A. The real image of an obstacle, through a converging lens, appears farther away than the object.

  B. The real or virtual image of an obstacle, through a converging mirror, appears farther away than the object.

  C. The real image of an obstacle, through a diverging lens, appears farther away than the object.

  D. The virtual image of an obstacle, through a diverging mirror, appears farther away than the object.

Answer: D

To answer this question, first eliminate answer choices (A) and (C), because we have a mirror, not a lens. Then draw ray diagrams for diverging (convex) and converging (concave) mirrors, and note that because the focal point of a diverging mirror is behind the surface, the image is smaller than the object. This creates the illusion that the object is farther away, and therefore (D) is the correct answer.

73. An object two meters tall is speeding toward a plane mirror at 10 m/s. What happens to the image as it nears the surface of the mirror?
(Rigorous) (Skill 19.2)

  A. It becomes inverted.

  B. The Doppler Effect must be considered.

  C. It remains two meters tall.

  D. It changes from a real image to a virtual image.

Answer: C

Note that the mirror is a plane mirror, so the image is always a virtual image of the same size as the object. If the mirror were concave, then the image would be inverted until the object came within the focal distance of the mirror. The Doppler Effect is not relevant here. Thus, the only possible answer is (C).

74. If an object is 20 cm from a convex lens whose focal length is 10 cm, the image is:
(Rigorous) (Skill 19.3)

    A. Virtual and upright

    B. Real and inverted

    C. Larger than the object

    D. Smaller than the object

Answer: B

To solve this problem, draw a lens diagram with the lens, focal length, and image size.

The ray from the top of the object straight to the lens is focused through the far focus point; the ray from the top of the object through the near focus goes straight through the lens; the ray from the top of the object through the center of the lens continues. These three meet to form the "top" of the image, which is therefore real and inverted. This is consistent only with answer (B).

75. A monochromatic ray of light passes from air to a thick slab of glass (n=1.41) at an angle of 45° from the normal. At what angle does it leave the air/glass interface?
(Rigorous) (Skill 19.4)

       A. 45°

       B. 30°

       C. 15°

       D. 55°

Answer: B

To solve this problem use Snell's Law:
$n_1 \sin\theta_1 = n_2 \sin\theta_2$ (where n_1 and n_2 are the indexes of refraction and θ_1 and θ_2 are the angles of incidence and refraction).

Then, since the index of refraction for air is 1.0, we deduce:
$$1 \sin 45° = 1.41 \sin x$$
$$x = \sin^{-1}((1/1.41) \sin 45°)$$
$$x = 30°$$

This is consistent only with answer (B). Also, note that you could eliminate answers (A) and (D) in any case, because the refracted light will have to bend at a smaller angle when entering glass.

76. A quantum of light energy is called a:
(Easy) (Skill 20.1)

       A. Dalton

       B. Photon

       C. Curie

       D. Heat Packet

Answer: B

The smallest "packet" (quantum) of light energy is a photon. "Heat Packet" does not have any relevant meaning, and while "Dalton" and "Curie" have other meanings, they are not connected to light. Therefore, only (B) is a correct answer.

77. Which statement best describes why population inversion is necessary for a laser to operate?
(Rigorous) (Skill 20.3)

   A. Population inversion prevents too many electrons from being excited into higher energy levels, thus preventing damage to the gain medium.

   B. Population inversion maintains a sufficient number of electrons in a higher energy state so as to allow a significant amount of stimulated emission.

   C. Population inversion prevents the laser from producing coherent light.

   D. Population inversion is not necessary for the operation of most lasers.

Answer: B

Population inversion is a state in which there are a larger number of electrons in a particular higher-energy excited state than in a particular lower-energy state. When perturbed by a passing photon, these electrons may then emit a photon of the same energy (frequency) and phase. This is the process of stimulated emission, which, when population inversion is obtained, can produce something of a "chain reaction," thus giving lasers their characteristically monochromatic and highly coherent light.

TEACHER CERTIFICATION STUDY GUIDE

78. **The constant of proportionality between the energy and the frequency of electromagnetic radiation is known as the:**
(Easy) (Skills 20.1 and 20.5)

  A. Rydberg constant

  B. Energy constant

  C. Planck constant

  D. Einstein constant

Answer: C

Planck estimated his constant to determine the ratio between energy and frequency of radiation. The Rydberg constant is used to find the wavelengths of the visible lines on the hydrogen spectrum.
The other options are not relevant options, and may not actually have physical meaning. Therefore, the only possible answer is (C).

79. The initial position and velocity of an electron (mass=9.1x10^{-31} kg) are measured. If the position is known to within a nuclear radius (10^{-14}m) and the velocity is known with the smallest accuracy allowed by the uncertainty principle, what will its position uncertainly be one year later? (Assume no force acts on the electron and h=6.6x10^{-34} J·s) *(Rigorous) (Skill 20.6)*

   A. 2.28 x 10^{18}m

   B. 7.27 x 10^{17}m

   C. 1.15 x 10^{10} m

   D. 3.3 x 10^{17} m

Answer: D

To solve this problem, we must use the equation expressing the uncertainty principle.

$$\Delta x \Delta p \geq \frac{h}{2\pi}$$

$$\Delta x \Delta (mv) \geq \frac{h}{2\pi}$$

We plug in the values from our problem:

$$(10^{-14}m)(9.1\times10^{-31}kg)\Delta(v) = \frac{6.6\times10^{-34}J\cdot s}{2\pi}$$

$$\Delta v = \frac{6.6\times10^{-34}J\cdot s}{2\pi(10^{-14}m)(9.1\times10^{-31}kg)} = 1.15\times10^{10}\,m/s$$

Now we consider its position after one year:

$$t\Delta v = (3.15\times10^{7}s)(1.15\times10^{10}\,m/s) = 3.3\times10^{17}\,m$$

Only answer D agrees with this; the other answers could result from arithmetic errors.

TEACHER CERTIFICATION STUDY GUIDE

80. Bohr's theory of the atom was the first to quantize:
    (Average Rigor) (Skill 21.1)

    A. Work

    B. Angular Momentum

    C. Torque

    D. Duality

Answer: B

Bohr was the first to quantize the angular momentum of electrons, as he combined Rutherford's planet-style model with his knowledge of emerging quantum theory. Recall that he derived a "quantum condition" for the single electron, requiring electrons to exist at specific energy levels.

81. Two neutral isotopes of a chemical element have the same numbers of:
    (Easy) (Skill 21.2)

    A. Electrons and Neutrons

    B. Electrons and Protons

    C. Protons and Neutrons

    D. Electrons, Neutrons, and Protons

Answer: B

To answer this question, recall that isotopes vary in their number of neutrons. (This fact alone eliminates answers (A), (C), and (D).) If you did not recall that fact, note that we are given that the two samples are of the same element, constraining the number of protons to be the same in each case. Then, use the fact that the samples are neutral, so the number of electrons must exactly balance the number of protons in each case. The only correct answer is thus (B).

82. When an electron is "orbiting" the nucleus in an atom, it is said to possess an intrinsic spin (spin angular momentum). How many values can this spin have in any given electron?
   (Average Rigor) (Skill 22.1)

   A. 1

   B. 2

   C. 3

   D. 8

Answer: B

To answer this question, recall that electrons fill orbitals in pairs, and the two electrons in any pair have opposite spin from one another. Thus, (B) is correct. Note that answer (D) is trying to mislead you into thinking of the number of valence electrons in an atom.

83. Which of the following statements best describes the diagram below?
   (Average Rigor) (Skill 22.1)

   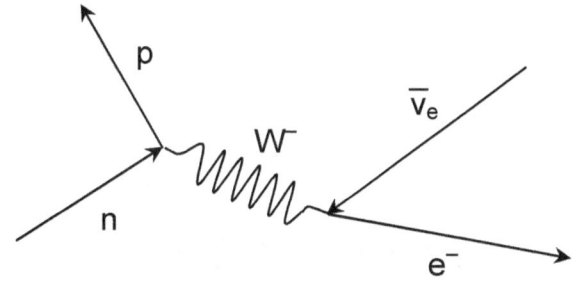

   A. A depiction of the mechanism for gamma ray emission

   B. A Feynman diagram of a photon exchange

   C. A Feynman diagram of a beta decay

   D. A depiction of the classical concept of electromagnetism

Answer: C

The diagram above is a Feynman diagram that depicts a neutron undergoing beta decay to produce a proton, an electron and an antineutrino. The W^- boson mediates the weak interaction in this case. Thus, the correct answer is C.

TEACHER CERTIFICATION STUDY GUIDE

84. Electrons are:
(Easy) (Skill 22.5)

- A. More massive than neutrons
- B. Positively charged
- C. Neutrally charged
- D. Negatively charged

Answer: D

Electrons are negatively charged particles that have a tiny mass compared to protons and neutrons. Thus, answer (D) is the only correct alternative.

85. When a radioactive material emits an alpha particle only, its atomic number will:
(Average Rigor) (Skill 23.1)

- A. Decrease
- B. Increase
- C. Remain unchanged
- D. Change randomly

Answer: A

To answer this question, recall that in alpha decay, a nucleus emits the equivalent of a Helium atom. This includes two protons, so the original material changes its atomic number by a decrease of two.

86. Ten grams of a sample of a radioactive material (half-life = 12 days) were stored for 48 days and re-weighed. The new mass of material was:
(Rigorous) (Skill 23.2)

    A. 1.25 g

    B. 2.5 g

    C. 0.83 g

    D. 0.625 g

Answer: D

To answer this question, note that 48 days is four half-lives for the material. Thus, the sample will degrade by half four times. At first, there are ten grams, then (after the first half-life) 5 g, then 2.5 g, then 1.25 g, and after the fourth half-life, there remains 0.625 g. You could also do the problem mathematically, by multiplying ten times $(½)^4$, i.e. ½ for each half-life elapsed.

87. In a nuclear pile, the control rods are composed of:
(Average Rigor) (Skill 23.4)

    A. Boron

    B. Einsteinium

    C. Isoptocarpine

    D. Phlogiston

Answer: A

Nuclear plants use control rods made of boron or cadmium, to absorb neutrons and maintain "critical" conditions in the reactor. However, if you did not know that, you could still eliminate choice (D), because "phlogiston" is the word for the imaginary element in an ancient structure of earth-air-water-fire.

TEACHER CERTIFICATION STUDY GUIDE

88. In a fission reactor, heavy water:
(Average Rigor) (Skill 23.4)

    A. Cools off neutrons to control temperature

    B. Moderates fission reactions

    C. Initiates the reaction chain

    D. Dissolves control rods

Answer: B

In a nuclear reactor, heavy water is made up of oxygen atoms with hydrogen atoms called 'deuterium,' which contain two neutrons each. This allows the water to slow down (moderate) the neutrons, without absorbing many of them. This is consistent only with answer (B).

89. **Which of the following pairs of elements are not found to fuse in the centers of stars?**
(Average Rigor) (Skill 23.5)

    A. Oxygen and Helium

    B. Carbon and Hydrogen

    C. Beryllium and Helium

    D. Cobalt and Hydrogen

Answer: D

To answer this question, recall that fusion is possible only when the final product has more binding energy than the reactants. Because binding energy peaks near a mass number of around 56, corresponding to Iron, any heavier elements would be unlikely to fuse in a typical star. (In very massive stars, there may be enough energy to fuse heavier elements.) Of all the listed elements, only Cobalt is heavier than iron, so answer (D) is correct.

90. Given the following values for the masses of a proton, a neutron and an alpha particle, what is the nuclear binding energy of an alpha particle?

   Proton mass = 1.6726 x 10^{-27} kg
   Neutron mass = 1.6749 x 10^{-27} kg
   Alpha particle mass = 6.6465 x 10^{-27} kg

(Rigorous) (Skill 23.6)

   A. 0 J

   B. 7.3417 x 10^{-27} J

   C. 4 J

   D. 4.3589 x 10^{-12} J

Answer: D

The nuclear binding energy is the amount of energy that is required to break the nucleus into its component nucleons. In this case, the binding energy of an alpha particle, which is composed of two protons and two neutrons, is calculated by first finding the difference between the sum of the masses of all the nucleons and the mass of the alpha particle. Using the equation $E = mc^2$ to find the energy in terms of the mass difference of 4.85 x 10^{-29} kg, and using the speed of light of about 2.9979 x 10^8 m/s, the result is the value given in answer D.

XAMonline, INC. 21 Orient Ave. Melrose, MA 02176
Toll Free number 800-509-4128
TO ORDER Fax 781-662-9268 OR www.XAMonline.com

<u>NEW YORK STATE TEACHER CERTIFICATION</u>
<u>EXAMINATION - NYSTCE - 2008</u>

PO#   Store/School:

Address 1:

Address 2 (Ship to other):

City, State Zip

Credit card number_____-_____-_____-_____   expiration_____
EMAIL _____
PHONE   **FAX**

ISBN	TITLE	Qty	Retail	Total
978-1-58197-660-1	NYSTCE ATS-W ASSESSMENT OF TEACHING SKILLS- WRITTEN 91			
978-1-58197-260-3	NYSTCE ATAS ASSESSMENT OF TEACHING ASSISTANT SKILLS 095			
978-1-58197-289-4	CST BIOLOGY 006			
978-1-58197-855-1	CST CHEMISTRY 007			
978-1-58197-865-0	CQST COMMUNICATION AND QUANTITATIVE SKILLS TEST 080			
978-1-58197-632-8	CST EARTH SCIENCE 008			
978-1-58197-267-2	CST ENGLISH 003			
978-1-58197-858-2	CST FRENCH SAMPLE TEST 012			
978-1-58197-344-0	LAST LIBERAL ARTS AND SCIENCE TEST 001			
978-1-58197-863-6	CST LIBRARY MEDIA SPECIALIST 074			
978-1-58197-623-6	CST LITERACY 065			
978-1-58197-296-2	CST MATH 004			
978-1-58197-290-0	CST MUTIPLE SUBJECTS 002			
978-1-58197-579-6	CST PHYSICAL EDUCATION 076			
978-1-58197-042-0	CST PHYSICS 009			
978-1-58197-265-8	CST SOCIAL STUDIES 005			
978-1-58197-396-9	CST SPANISH 020			
978-1-58197-258-0	CST STUDENTS WITH DISABILITIES 060			
			SUBTOTAL	
	FOR PRODUCT PRICES VISIT **WWW.XAMONLINE.COM**		**Ship**	$8.25
			TOTAL	

www.ingramcontent.com/pod-product-compliance
Lightning Source LLC
Chambersburg PA
CBHW080534300426
44111CB00017B/2727